WINDING THROUGH TIME

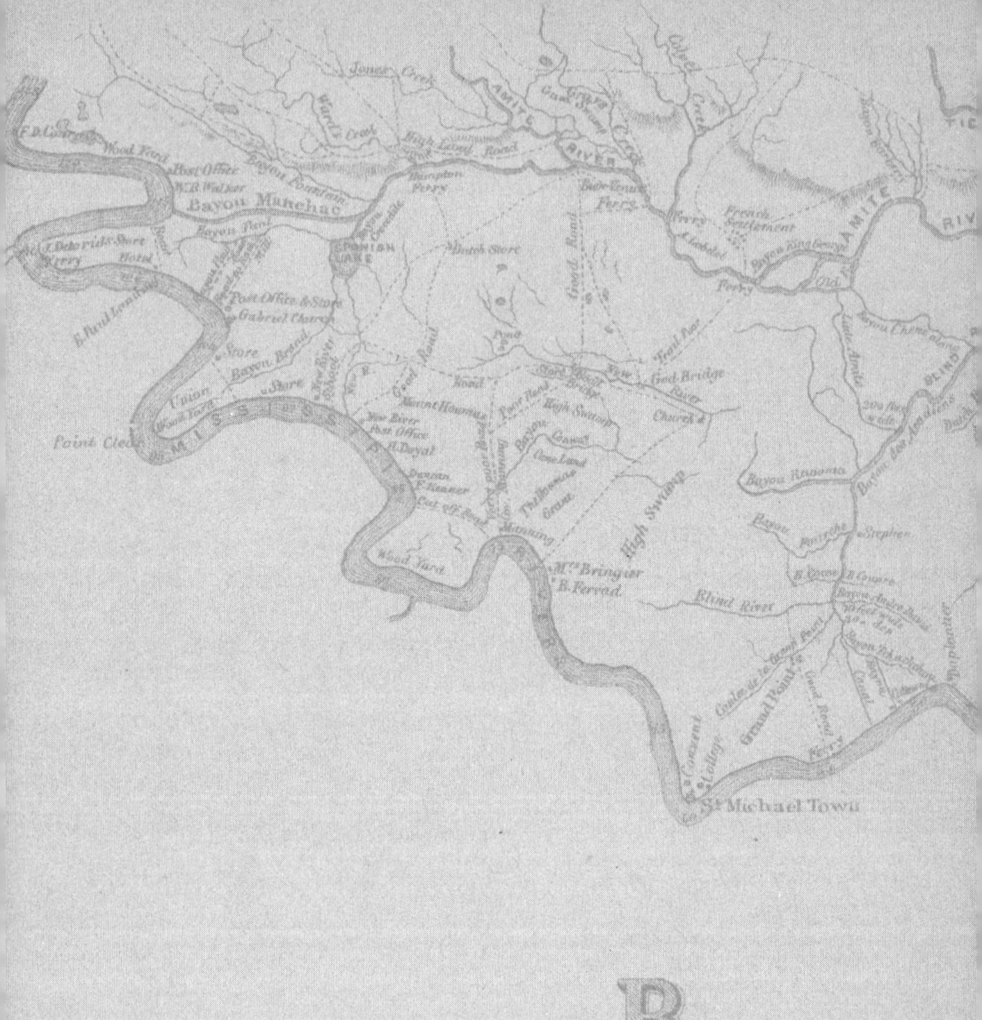

B.

Extract from Dep.t of the Gulf Maps

Prepared under direction of

B.VT. B.RIG. G.ENL. M. D. M.c ALESTER, M.

To accompany his Report on his Examination and

PASS & BAYOU MANCHAC & AMITE

SCALE *of Miles.*

6 0 6 12

To Pam and Charles and Jonathan,
with thanks for sharing their bayou with me.

Contents

WINDING THROUGH TIME

The Forgotten History and Present-Day Peril of Bayou Manchac

MARY ANN STERNBERG

LOUISIANA STATE UNIVERSITY PRESS ⚜ BATON ROUGE

Published by Louisiana State University Press

Manufactured in the United States of America
First printing

Designer: Barbara Neely Bourgoyne
Typeface: Adobe Jensen Pro
Printer and binder: Edwards Brothers, Inc.

Library of Congress Cataloging-in-Publication Data
Sternberg, Mary Ann.
Winding through time : the forgotten history and present-day peril of Bayou Manchac / Mary Ann Sternberg.
p. cm.
Includes bibliographical references.
ISBN 978-0-8071-3253-1 (cloth : alk. paper) 1. Manchac, Bayou (La.)—History. 2. Manchac, Bayou, Region (La.)—History. 3. Manchac, Bayou (La.)—Environmental conditions. 4. Manchac, Bayou, Region (La.)—Environmental conditions. I. Title.
F377.M23S74 2007
976.3'3—dc22
2006039054

Map details on pp. ii–iii and 130 from J. K. Hezlep's 1868 map of the Bayou Manchac area, in the Collection of the Louisiana State Museum. Details on pp. xiv, 38, 102, and 148 from George Gauld's 1774 *A Plan of Manchac,* LSU Libraries' Special Collections. Detail on p. 10 from a 1779 map by Joseph DeBarres, LSU Libraries' Special Collections. Detail on p. 92 from George Gauld's 1779 map of West Florida, LSU Libraries' Special Collections.

The paper in this book meets the guidelines for permanence and durability of the Committee on Production Guidelines for Book Longevity of the Council on Library Resources. ♾

Maps

Illustrations

WINDING THROUGH TIME

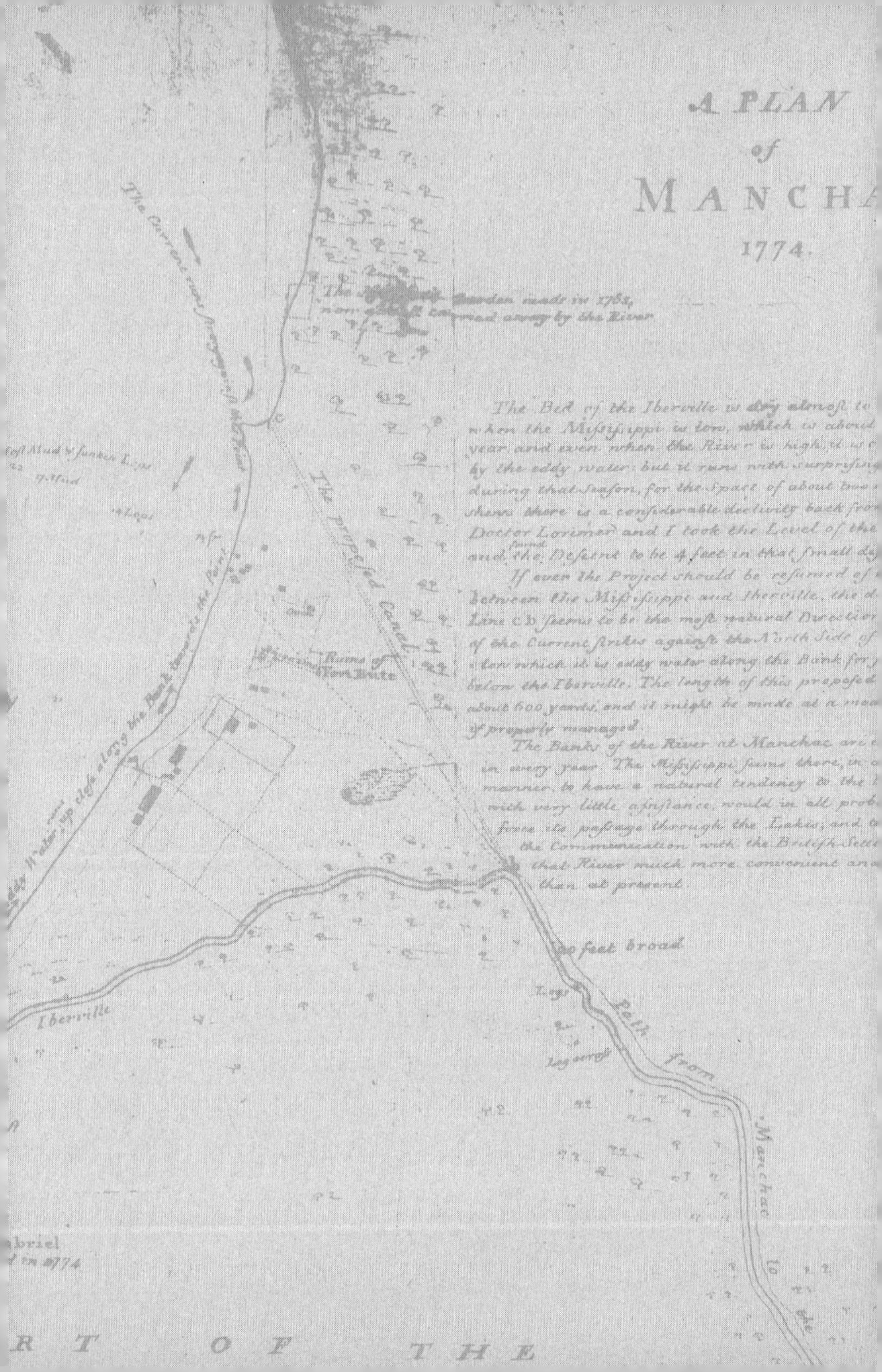
A PLAN
of
MANCHA
1774.
now
carried away by the River
The Bed of the Iberville is dry almost to
when the Missisippi is low, which is about
year, and even when the River is high, it is
by the eddy water: but it runs with surprising
during that Season, for the Space of about two
shews there is a considerable declivity back from
Doctor Lorimer and I took the Level of the
and found the Descent to be 4 feet in that small da
If ever the Project should be resumed of
between the Missisippi and Iberville, the d
Line C D seems to be the most natural Direction
of the Current strikes against the North Side of
Iberville, below which it is eddy water along the Bank for
below the Iberville. The length of this proposed
about 600 yards, and it might be made at a mod
if properly managed.
The Banks of the River at Manchac are
in every year. The Missisippi seems there, in a
manner, to have a natural tendency to the
with very little assistance, would in all prob
force its passage through the Lakes; and
the Communication with the British Settle
that River much more convenient and
than at present.
The Current runs from the Point
The Proposed Canal
Ruins of Fort Bute
Eddy Water runs up close along the Bank towards the Point
Iberville
20 feet broad
Path from Manchac
R T O F T H E

Introduction

I FIRST NOTICED THE BRASS MARKER canted out of the grassy base of the broad Mississippi River levee about six years ago as I drove slowly along the River Road south of Baton Rouge, Louisiana's capital city. The narrow, winding asphalt ribbon follows the river, which flows unseen behind the hulking form of the levee.

The historical marker read:

> Bayou Manchac. (Iberville River). Iberville returned from the Mississippi River to his fleet at Ship Island via this bayou in 1699. Boundary: Spanish Isle of Orleans and British West Florida, 1763–1783, and Territory of Orleans and Spanish District of Baton Rouge, 1803–1810.

Intrigued, I stopped to investigate this place of sketchy description that seemed to have disappeared. Instead of a bayou or a river, there was only a small lagoon on the side of the road, a pool of black water ringed by a stand of cypress trees, barely visible and hardly navigable.

I knew of Bayou Manchac only vaguely—that it was once a distributary* of the Mississippi and had been cut off from the river

* A distributary is a natural outlet from the river that carries off spring floodwaters into the low backlands.

when the Corps of Engineers erected the monumental levee all along its banks following the great flood of 1927. The levee might have explained the lack of water in the channel except that I was familiar with two other distributaries located nearby, Bayous Plaquemine and Lafourche, that wind from the west bank of the Mississippi and had also been leveed at the river. They seemed to have remained much more viable waterways.

In 2001, Manchac had flooded exceedingly in the aftermath of Tropical Storm Allison and had been the subject of dramatic local news that showed fine brick homes, identified as "near Bayou Manchac." They were isolated in a lake of swirling water, their partly submerged windows looking like wide-eyed, panicky swimmers. But I couldn't identify where these houses might be because I didn't know where the bayou went.

Bayou Manchac was, in short, a mystery.

So I harbored minimal knowledge and little curiosity about

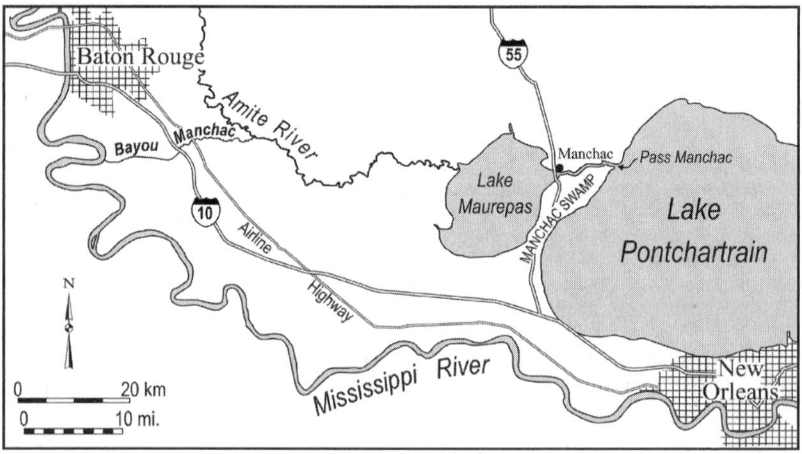

Interstate 10 and Airline Highway, major highways between Baton Rouge and New Orleans, traverse Bayou Manchac.

the waterway until the celebration of the bicentennial of the Louisiana Purchase loomed in 2003, when an interesting but esoteric fact struck me: the bayou had been a boundary of that magnificent land deal. It had been one of the waterways that defined the Isle of Orleans, Thomas Jefferson's original, and primary, target when he initiated negotiations with France.

The national celebration of the Purchase, however, focused on Lewis and Clark's exploration of the great and unknown western territory gained in 1803 rather than on the familiar city of New Orleans and its environs. But I convinced a regional magazine that this small, now-forgotten bayou might make a timely and relevant subject for an article.

As I researched Bayou Manchac, I became fascinated and increasingly convinced that its story was much better than I'd suspected—more colorful, richer, and much more interesting. And I also realized that the evolution of this little waterway, though unique, is a cautionary tale for other small historic American streams.

So I followed the thread, enticed by people like Pam Caillouet.

Pam, an intelligent, energetic, no-nonsense woman, and her husband, C.R., a nationally recognized high-definition video technologist, have lived on the bayou since 1993. Their A-frame log cottage, raised high on timbers as sturdy as telephone poles, overlooks the stream. The tall posts were added, she told me, by the previous owner after the piers of his original house proved inadequate: the house flooded shortly after he moved in.

A broad deck spans the Caillouets' living area under a canopy of trees. It is festooned with bird feeders that have attracted bluebirds, painted buntings, and goldfinch, as well as a myriad of more common birds. From this aerie, they delight in the parade of animals that cross their property—rabbits, raccoons, possums, the occasional red fox. They watch wood ducks and turkey landing nearby, otters scrambling near their bank, and alligators gliding slowly past in the bayou.

The Caillouets and many of their neighbors live within their landscape, in tune with the changing stream that borders their property. They observe the seasons, heralded by variations in the light that dances on the water, and share sightings of the earliest spring iris and migrating birds. And they have, by necessity, learned the secrets of living in harmony with the sometimes unruly waterway, having become savvy about such things as where to move their cars when high water is predicted.

But what really captivated me was a casual remark Pam made when we first talked about the bayou. "I can feel the ghosts," she said in a quite rational tone, somehow making me feel that the collective spirits of those who had lived or traveled along Bayou Manchac might linger somehow—a parade of prehistoric natives, Iberville, French *voyageurs,* British and Spanish colonials, trappers and Civil War soldiers, steamboat captains, frontiersmen and farmers, loggers, fishermen, and all those who have come into contact with this waterway over the centuries.

Pam and C.R. and others like them have also watched a bit warily as new development has crept along the bayou banks, with many of the newcomers paying but scant attention to the bayou's history or its beauty.

As I have become more familiar with Bayou Manchac, I too can occasionally feel the ghosts. It is irrational, I know, and yet it will partially explain how I happened to write this book: somehow, it seemed as if Bayou Manchac chose me to tell its story, which is why this became more than just a straightforward history.

1 What Is Manchac?

LOUISIANA'S RIVERS AND BAYOUS have been integral forces in the history and development of the state, as water transport was the most efficient way to move goods and people over any distance. In verdant south Louisiana, much of the land was swampy, covered with dense overgrowth, or mucky underfoot. Its waterways served as not only arteries of transportation but also as the sites of villages and sources of food.

Bayous are, by definition, a particular kind of stream, usually narrow, winding, barely moving. Some are black-water meanders

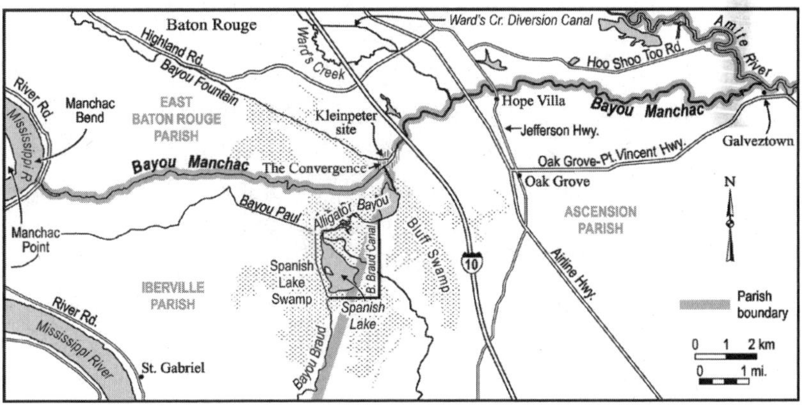

The bayou and nearby landmarks today

through pristine landscapes; others resemble water highways, dynamic with boat traffic, their banks lined with houses, docks, and businesses, their channels spanned by bridges. The character of Bayou Manchac lies somewhere between.

Like many other place names in Louisiana, the word *Manchac* is derived, at least indirectly, from a Native American language. The French, first to explore and settle the area, adapted it from *imashaka,* the combination of two Choctaw words—*im,* roughly translated as "behind it," and *šaka* for "rear."* The natives may have also referred to the bayou as "Ascantia."

What became known as Bayou Manchac was a narrow waterway that offered the first link in a chain of waterways that comprised a "back route" linking the Mississippi River with the Gulf of Mexico. This passage followed Manchac, the Amite River, and Lakes Pontchartrain and Maurepas to the Gulf, significantly shortening the distance, by approximately 125 miles, rather than following the meandering Mississippi down to its mouth. This back route would also have facilitated a part of the upriver journey before steam-powered boats conquered the formidable Mississippi River current.

Indian natives shared the existence of this back route with Pierre Le Moyne, Sieur d'Iberville, who, with his brother Jean-Baptiste Le Moyne, Sieur de Bienville, had been dispatched by French king Louis XIV to chart the mouth of the Mississippi. In Iberville's journal of his 1699 expedition is the first written documentation of the bayou's existence. Shortly thereafter the little

* According to linguistic scholar William Bright, in his article "Native American Place Names in the Louisiana Purchase."

stream became a prominent feature on maps of the Louisiana territory, beginning with Guillame De l'Isle's 1702 map, "Carte de la Louisiane et cours du Mississipi."

Early colonial maps labeled the thin line winding between the Mississippi and the Amite rivers as the Iberville River, variously spelled "Ibberville" and "Ibbeville," or Manchac. (During the British colonial period, *Iberville River* also often referred to Bayou Manchac and the lower Amite River together.) Occasionally the stream was identified as Ascantia, or a misspelled version, Akankia, or was translated as Manchacq or Manchaque, Mashake Creek, or Massiac. But it invariably appeared on maps of the territory.

The name Manchac must have greatly appealed to Europeans' sensibilities, however, because they applied it liberally. Manchac was the designation of both British and Spanish governmental jurisdictions during the colonial period—the Manchac District, and the District of Manchac or Iberville. A tiny British settlement named Manchac was tenuously planted in 1763 on the upper bank of the bayou at its junction with the Mississippi and existed for thirty years.

It was replaced in the mid-nineteenth century by a namesake settlement approximately a mile and a half north of the previous community. This one boasted a post office, a railway depot, and, by 1895, a population of 126. But it met with destruction during the early twentieth century, undermined by a great spring flood.

A Spanish post established in 1767 on the downriver bank of the bayou at its junction with the Mississippi was called Fort San Gabriel de Manchac. The Catholic church nearby, completed in 1776, was first known as the Manchac Church or San Gabriel of Manchac, or Manchak, or Manchack.

In 1805, a mail stage between New Orleans and St. Francisville (north of Baton Rouge) stopped at Manchac, which was neither the early British settlement nor the later American village. The stage passed through the downriver community where the Spanish church was located, the predecessor village to what is now the small town of St. Gabriel.

The name *Manchac* also designates Mississippi River landmarks, including Manchac Bend—the arcing river curve from which the bayou begins—and Manchac Point, the long peninsula opposite. In the antebellum era, Manchac Plantation was a great sugar estate located on the downriver junction of the bayou and the Mississippi. It has been immortalized on contemporary Army Corps of Engineers maps that still identify the riverbank there as "Manchac Landing."

Today, these references are obscure, and mention of Manchac or Bayou Manchac almost always elicits a blank response. It is a tiny and insignificant thread of blue found only on the most detailed local maps. Or there is instant recognition—for another, better-known place with the identical name.

This confusion of place names is not new. An 1877 listing of Louisiana boat landings identified one landing named Manchac on the Mississippi River and another between Lakes Pontchartrain and Maurepas. It is this latter Manchac that has instant identity today.

Here, a tiny community named Manchac (also known as Akers) is perched at the end of a seven-mile strait called Pass Manchac that connects Lakes Maurepas and Pontchartrain. The pass is surrounded by the expansive, well-known Manchac Swamp. Interstate 55 cuts through this area and highlights "Manchac" on a large green exit sign.

The bayou called Manchac, however, is not located here; it is approximately thirty-five miles away as the crow flies and much farther over land or by water.

Two major highways and four smaller roadways cross the bayou's channel, and several rural roads run parallel. The older of the large highways, U.S. 61, known locally as Airline Highway, dates from 1935 and once served as the main artery between New Orleans and Baton Rouge. Its Manchac crossing has been decorated with an historical marker since 1960, similar to the one that initially intrigued me, but the small plaque is little noted by drivers or their passengers as they speed past. The other major road, Interstate 10, carries more than 55,000 vehicles across Bayou Manchac daily, but they have no reason at all to distinguish this narrow, unidentified, tree-shaded stream from any other that flows beneath the highway between Jacksonville, Florida, and Los Angeles, California.

Bayou Manchac runs through the landscape just as it did when it was one of the integral forces that formed Louisiana. But now it is a place of history, utility, and beauty, most often overlooked, misunderstood, or threatened.

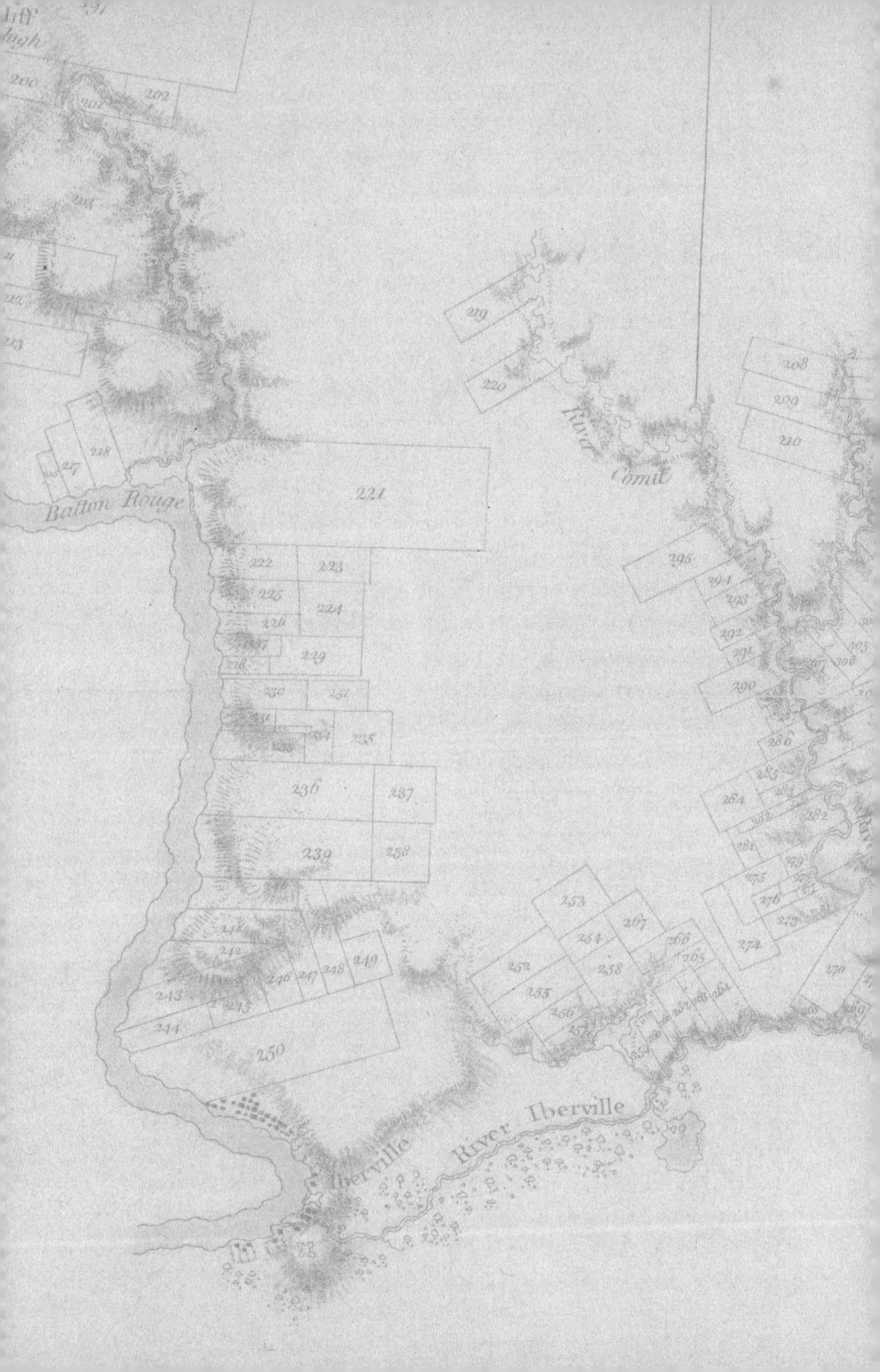
Batton Rouge
River
Comit
River Iberville
Iberville
221
222
223
224
225
226
229
230
235
236
237
239
241
242
243
244
245
246
247
248
249
250
252
253
254
258
265
266
267
274
276
284
285
286
290
291
292
293
294
295
208
209
210
220
218
217
200
201
202

2 Exploring the Bayou

I WANTED TO SEE THE BAYOU from the inside before delving into its history, so I found a genial young man with a small motorboat. He was a local resident who had often fished Manchac with his father. It was a place he liked going, he confessed, but they had never caught many fish there.

We entered from the mouth, on the bayou's eastern end at the Amite River, and crossed a surprisingly distinctive shredded line that marks where the brisk, toffee-colored flow of the larger Amite accepts the clear, languid, tea-colored water of Bayou Manchac. The bayou channel seemed confined and intimate between tree-lined banks; the hum of the little outboard sang baritone to a chorus of buzzing insects. Swirls of emerald green duckweed and tiny fire-red trumpet flowers bobbed on the flat water like decorations in a punch bowl. Occasionally ripples broke the water as a fish surfaced. Once, we passed a flotilla of young mullet swimming together, their pointed heads above water and moving in synchronized motion like a comedic water ballet.

The bayou in places was strikingly, surprisingly beautiful. One long reach was lined with towering cypress, handsome live oaks, sycamore, walnut, ash, and magnolia that formed a green archway. Ropey, heavy-leaved vines dangled to the water like exotically beaded curtains. Along the low mud banks, a forest of palmettos waving spiky hands stood amid clusters of ruddy cypress knees that rose as elegant and artful from the mud as the modeled sculptures of Henry Moore. Elsewhere, the bank was shrouded with stands of cane, as thick and impenetrable as early explorers' accounts described them.

A gray and white belted kingfisher chittered and darted along the bayou ahead of us like a scout, landing briefly on the limb of a large tree fallen across the channel. The bird seemed oblivious to the raft of debris—white Styrofoam cups, blue plastic jugs, slashed black garbage bags—that bobbed against the upstream side of his perch.

In the kingfisher's wake, the whispery, white flutter of a snowy egret moved the still air and, from somewhere, a barred owl committed a series of melancholy hoots.

No bridge crossed the channel in the first five miles west of the Amite save a span of silver filament that reached from bank to bank across the chasm of more than fifty feet. In its center hung an intricately woven spiderweb, its creator clinging to the laciest inner inch of the square netting. As a soft breeze rustled the web, this engineering marvel swayed on its axis, tilting gracefully, primed to snare the next hapless visitor.

Such stretches of residual wilderness beauty make the bayou seem removed from civilization. Yet just around the next gentle bend might be a string of wooden shacks hard by trailers raised on cinder blocks and rustic camps with antennae sprouting from their roofs with dilapidated piers crumbling into the water. Or there may be a development of elegant houses and landscaped gardens with an incongruous black-and-white sign decorating their view of the bayou: PIPELINE: DO NOT DREDGE.

The few residents visible were as diverse a group as the buildings. A red-haired boy of eight or nine skidded his dirt bike down the steep mud bank and huffed back up again; an elderly, white-haired woman, reduced to a great amorphous, white-sheeted form, endured an outdoor haircut; a clean-shaved executive type in a crisp, pink, enjoy-the-weekend polo shirt leaned from a well-tended wooden dock to scowl at his limp fishing line; a rough-looking, bare-chested man in green camouflage cutoffs leaned under the hood of a battered truck in a yard filled with machine parts. Along several miles of the north bank, the neighbors included small herds of Holsteins and Herefords clustered at tree-shaded fence lines, grazing laconically, observing our passage.

The first bridge across the channel rose on stork legs, rumbling with car and truck traffic. Only this crossing—the old Jefferson Highway bridge—and a weathered railroad trestle nearby were elevated across the bayou. The other roadways run flat, at road grade bank to bank, clearly designed with no consideration for boat traffic.

Ten miles west of the Amite River entrance, Bayou Manchac abruptly dead-ends. A great swath of overgrowth hides the ditch that Manchac becomes. Beyond this point, the Manchac channel is mostly dry, its bed a wild trough of trees and shrubs.

As we passed beneath the Jefferson Highway bridge, I noticed a modernistic, matte-finished metallic cube, flecked with rust-colored decorations, sunk on an angle into the bank of the bayou. At a museum of contemporary art, I might have felt compelled to scrutinize this piece for its deeper meaning, but along the Manchac, it was only an unwanted stove that someone had hauled to the bank edge and unceremoniously dumped.

Such assaults on the landscape, I learned, were common. On subsequent trips, I began to keep a list of "bayou artworks," which grew to include a crushed dishwasher, a heap of roofing material, a four-burner stove top, a doorless refrigerator, the frame of a plow, a bicycle carcass, car fenders, and too much more.

And yet, on that first outing, not far from the half-sunken stove, I spied a pair of wild turkeys on the crest of a ten-foot bluff, their large, mottled brown bodies poised like statues beneath a great beckoning live oak and beside a graceful wrought-iron streetlamp. Iberville himself had noted turkeys when he passed this way. The turkeys were sentinels for a new residential subdivision, its streets already laid, awaiting construction of new homes.

The addition of houses along Bayou Manchac could be welcome, but many of the new developments are designed so that the rear of each house is separated from the bayou with a fence, giving the impression that the waterway is not important enough to invite their attention. And I wonder: as more and more such developments crowd the banks, what will become of the turkeys?

3 Community Development, Then and Now

STEVE FULLEN HAS CLEARED a tabletop amid the jumble and clutter of the expansive archaeology lab in a basement room on the LSU campus. Fullen, collections manager for the anthropology section of the LSU Museum of Natural Science, has pulled a plain corrugated cardboard box from storage, from which he extracts several Ziploc bags. He scatters their contents into a composed design of smooth, dusty, jagged fragments in muted tones of ochre, gray, and rust. Some are smaller than a pencil eraser, others as large as the heel of my hand. One item resembles a short, slightly curved white stick; another is a plain mustard-colored rock; still another is a rough, dark-brown pyramid-shaped clod.

It looks like rubble.

It *is* rubble, but with an impressive provenance. These undistinguished-looking pieces are artifacts excavated from the Kleinpeter site, located near the junction of Bayou Manchac and Bayou Fountain, considered one of the most significant prehistoric archaeological finds in the region. Archaeologists believe that successive native cultures inhabited the site for almost two millennia, from approximately 250 B.C. until just prior to Iberville's passage.

The location would have been almost ideal for native settlement, offering streams for transportation, level ground raised above the floodplain, rich soils for farming, and bountiful hunting. The tract sloped down to both Bayou Fountain and Bayou Manchac and was probably surrounded by an expanse of rolling green land studded with magnificent stands of trees and dense overgrowth. Alligator Bayou entered Manchac from the south nearby, forming a confluence of three waterways that became known as the Convergence.

Iberville had remarked on this place in his journal: "Here two small streams unite with this one [Manchac], one from the north-northwest and the other from the southeast, making it half again bigger and twice as wide.... [It is] One of the prettiest spots I have seen, fine level ground, beautiful woods, clear and bare of canes...."* He also wrote of seeing six pirogues (dugout canoes) there, but he failed to mention seeing people, a settlement, or mounds. This omission has suggested to historians that the last of the native communities had already moved away by 1699, since Iberville described in extraordinary detail a visit he had made just ten days before to a large Bayou Goula village along the Mississippi River.

Archaeological evidence indicates that the last occupants of the Kleinpeter site may have been similar to the natives Iberville encountered at the Bayou Goula village and to those encountered by French explorers who followed Iberville. They wrote about the sophistication of the natives' lifestyles, surprised that although the Indians lacked writing and domesticated animals, they had their own spoken language, ritual celebrations, and forms of games and recreation. They used a variety of tools and weapons and fashioned pottery and basketry decorated far beyond mere utility.

The Kleinpeter artifacts revealed that the prehistoric people who lived at the Convergence hunted the nearby forests and swamps and fished the streams. They made tools and ornaments

* References to Iberville's journals throughout this book are from *Iberville's Gulf Journals*, translated and edited by Gaillard McWilliams (University of Alabama Press, 1981).

from bones, antlers, and animal teeth and created clothing and blankets from deer and bear hides. They hunted for food—deer, bear, buffalo, raccoons, opossums, squirrel, muskrat, rabbits, foxes, turkey, geese, duck, and egret—and fished for gar, catfish, buffalo fish, bowfin, freshwater drum, alligator, and turtles with nets and spears. Iberville had noted the area's prolific wildlife, particularly the fish and "crocodiles" in the bayous.* The Convergence was so rich, in fact, that later natives called it Anamatah, "the fish place," a reputation it sustained until the 1950s.

Edible wild plants were bountiful—varieties of berries and nuts, grapes, persimmons, wild onions, freshwater lotus seeds, wild potato and other tubers. Later native cultures cultivated maize and perhaps beans, melons, sunflowers, tobacco, and squash. They used the thick cane growing along stream banks to make poles for construction and sharp tools and created pirogues by hollowing fallen cypress trees with fire. For their era, the prehistoric occupants of the Kleinpeter site led a remarkably rich life.

The exact date when natives decamped from the peninsula at Bayous Fountain and Manchac is unknown. They had certainly moved away by the time of Iberville's passage; however, when British mapmaker Elias Durnford noted an "Indian camp" on his 1770 map, he gave no indication that the site was uninhabited. Yet when the property first appeared in the British colonial land register of West Florida in 1778, it was listed as "uncultivated plantation property and pasture," suggesting that the cultural treasure of mounds and artifacts tucked into its back corner had not been discovered.

No records indicate that anyone lived on or near the archaeological site after the natives departed save John McKeough, an

* Iberville was only the first of many visitors to the area who would confuse the American alligator with its cousin the crocodile, which is not indigenous to south Louisiana.

Irish farmer who held two land grants along the bayou. He may also have cut and sold cypress and operated a primitive ferry across the streams. But McKeough seems not to have discovered the mounds or their treasures because, when he died in 1804, his estate included no native artifacts. It did, however, list among his possessions "1 rat trap; 2 cow hides and candlesticks; 2 muskito-bars; 1 log chain; a grindstone; a lott of Indian corn; a pott, skilett, iron lid and griddle," six slaves, and seven horses.

The existence of several prehistoric mound sites along Bayou Manchac and its tributaries was first revealed in the late nineteenth century, and the six mounds that comprised the Kleinpeter site were quickly identified as the most important, most dramatic, and best preserved. In 1938, cultural geographer Fred Kniffen first documented the Kleinpeter earthworks, although the first photographs did not appear publicly until 1965. Through the years, amateur archaeologists removed exposed artifacts but, remarkably, the mounds were left virtually undisturbed for two centuries, obscured by heavy overgrowth and accessible only from the bayous.

This changed abruptly in 1985. The Kleinpeter site was part of a vast suburban parcel slated to become the first exclusive residential golf community in Louisiana's capital city. The Country Club of Louisiana would be a gated enclave of elegant homes and shady lanes winding around a Jack Nicklaus signature golf course; its lucky residents would be the first to make their homes in this idyllic, never-before-occupied spot. Or so the brochures might have claimed until bulldozers began to clear the land and rip away the overgrowth, revealing the prehistoric mounds rising as distinctly and unnaturally from the landscape as the new bunkers being sculpted onto the golf course.

The potential destruction of this significant prehistoric treasure rallied the traditionally low-profile archaeological community, who urged developers to set aside the tract, to preserve it as a park, or at least to refrain from improving the lots on which the mounds were located. But no regulations existed to enforce their requests. The developers were obligated, however, to document the wetlands on their property, and so a team of archaeologists was contracted to make a cursory exploration. Then the gentle curves of Muirfield Circle were laid out to twine among the mounds, and lots were staked. The matter was settled.

Several years later, perhaps lured by the ghosts of the bayou or merely by serendipity, a local businessman named Ben Kleinpeter enrolled in a short course on archaeology at Louisiana State University. Professional archaeologists Malcolm Shuman and Dennis Jones introduced the class to numerous prehistoric treasures throughout the state, including the important Kleinpeter site, and were surprised and delighted when their student approached them, identifying this significant place as property that had once belonged to his family.

"We used to squirrel hunt and coon hunt there," Ben Kleinpeter said, remembering boyhood outings during the 1930s and '40s. They would arrive by boat from Bayou Fountain or Bayou Manchac at this "thick and primitive area. There was a lot of superstition about the place," he chuckled, "and a lot of talk about panthers."

Inspired by learning of the site's importance, in 1990 Kleinpeter enlisted the support of the Baton Rouge Area Foundation to raise sufficient funding to underwrite an extensive excavation of the mounds. "I couldn't wait for them to get out there," he said, reveling in the thrill of the project many years later. "We were always curious about [that place]."

Dennis Jones, Malcolm Shuman, and two associates viewed the contract as a prized assignment, Shuman confessed. "Touching artifacts over two thousand years old was exciting." The natives had been there for many generations, he pointed out, and had created "their own world view."

But his team wasn't seeking to uncover the ancient inhabitants' world view. They began to make a scientific analysis of the culture and behavior of the unknown peoples who had lived at the junction of Manchac and Bayou Fountain but left no written records. By carefully studying the layout and found objects—stone tools, ceramic pieces, bone fragments, fish scales, teeth, arrowheads, and more—the archaeologists could attempt to reconstruct details of daily life of those first inhabitants of the back corner of the land that is now Country Club of Louisiana.

The team plotted the six mounds, which ranged from small (just under two and a half feet high) to impressive (nine and a half feet high). They designated four mounds on the bayou side of the street B, C, D, and E. The remaining two, on the other side of the street, were dubbed A and F and were on properties whose owners did not wish to participate in the project.*

The team's analysis revealed that the mounds had been constructed about 1100 A.D., positioned around a large plaza approximately the size of three basketball courts. Homesites surrounded the perimeter; houses or ceremonial structures rested on top of the larger mounds. Given the scale of the place, Dennis Jones surmised, the Kleinpeter site must have been a very important regional center in 1300 A.D. or so, "perhaps a provincial capital, where natives from around the area would gather."

* A diagram showing the mound locations is included with the illustrations following p. 34.

Clustered around the mounds were middens—stashes of refuse, broken, no-longer-useful detritus of life, discarded by residents year after year that were the sources of tangible clues to the real lives of these unknown people—artifacts that filled fifty cardboard boxes and found their way to a lab on the LSU campus.

Steve Fullen wore an expression of open delight as he lightly fingered the odd bits and pieces on the table, lapsing warmly into archaeological vernacular. "Ceramic sherds" he called the fragments of pottery containers, and "lithic debitage" the tools created from worked stone. There were "daub fragments," bits of material such as those from a constructed oven, and "worked bone," animal pieces made into tools. As he touched them and explained their uses and meanings, the value of the unprepossessing rubble leapt across centuries, becoming real cooking pots and real working tools of people who lived in a certain place along Bayou Manchac centuries ago.

One ragged fragment is the incised decorative lip of a plate. Another, the rim of an expertly formed bowl, its sophistication revealed by the density of the clay mix visible within the broken edge. Another, Fullen guessed, would have been part of the curved neck of a flask.

He flipped over a large, imperceptibly curved sherd to reveal on its underside a coarsely scratched design of angled horizontal and vertical lines. "Mazique incised," he intoned, a style dating from the Tchefuncte period, 250 B.C. to 200 A.D. Another sherd, shaped like a rounded corner, was perforated with a tight, sophisticated lattice design. "Churupa punctated," Fullen suggested, "probably a decorative scallop from the rim of a large bowl" from the Troyville period.*

The curved white stick was actually a bleached deer bone fashioned into a hole punch, probably used for sewing; the roughly

* Native American cultural periods are in a table in the endnotes.

made, triangular clod seemed to be the foot of an early cooking pot, burned to its opaque hue from frequent exposure to pit fires. The mundane-looking rock was, on closer examination, a tool with evenly chipped, sharpened edges. A scraper, Fullen offered, perhaps to clean hides or remove meat from the shells of large turtles, like ones I'd seen along the bayou.

The Kleinpeter site's artifacts, analyzed documented, and protected, link the first residents of this land with those who have come there most recently. The mounds, however, though still rising unnaturally from the landscape, exist at the whim of the landowners on whose property they are located.

Perhaps the ghosts of the Tchefuncte, the Coles Creek, or the Plaquemine cultures will have influence enough to ensure that the mounds remain unmolested. If not, these prehistoric constructions may be used as so many other Louisiana Indian mounds have been—dug into for landfill, roadbeds, and gardens—and disappear forever.

4 Iberville Discovers the Bayou

IN 1698, PIERRE LE MOYNE, Sieur d'Iberville was summoned to Paris by Lord Pontchartrain, Minister of the Marine, who had chosen the naval hero of New France (Canada) to lead an expedition to the Gulf of Mexico, to locate the mouth of the Mississippi River As Iberville wrote in his journal, he was to "select a good site that can be defended with a few men, and block entry to the river by other nations." By this defensive strategy, France would secure its territorial claim on the Mississippi River valley before its rivals could contest it.

During the late seventeenth century, Great Britain, Spain, and France were all actively claiming and colonizing portions of the New World. The British had settled the east coast and parts of Canada; the Spanish claimed Florida and what is now the American southwest. But although French explorer Robert Cavelier, Sieur de La Salle, had claimed the Mississippi River valley in 1682, he had traveled from the upper river only as far south as what is today known as Head of Passes, where the river splays into finger channels toward the Gulf. There he had planted a cross and claimed the river valley for France and King Louis XIV. But La Salle had never actually reached the Gulf of Mexico or entered the river from the south. So to establish themselves against British expansion and

Spanish incursion, the French felt compelled to aggressively define their colony to the Gulf and also begin its settlement.

The thirty-eight-year-old Quebecois has been described by an admiring biographer as one whose "achievements and feats of endurance [were] so numerous and so extraordinary that if they were not well documented, most would find it hard to believe they ever happened." Iberville was brave and heroic and, according to some accounts, also quite ruthless.

With his younger brother Jean-Baptist and a crew of several hundred men, Pierre Le Moyne set forth from the West Indies in four sailing barks. They approached Pensacola, Florida, and were rebuffed by the Spanish, then landed near present-day Biloxi, Mississippi, where Iberville ordered the ships' cannons fired to attract attention. Among the curious to appear was a Bayou Goula chief who happened to be in the area with a buffalo hunting party. They had traveled from their home on the west bank of the Malbanchya, as the natives then called the Mississippi River. The chief told Iberville of the large river to the southwest, which encouraged the explorers to continue west along the coast. On February 27, 1699, they were rewarded with discovery of the true mouth of the Mississippi.

Iberville and his party sailed up the river, stopping at the village of the Bayou Goula, who then shared their land with a remnant population of Mugulasha. After hosting the French adventurers, the Bayou Goula chief and a guide joined the expedition and accompanied them upriver to another native village. Here, the Indians showed Iberville a letter written by Henri de Tonti, who had been a member of La Salle's 1682 expedition. This proved that the great river on which they now sailed was indeed the Mississippi River of previous exploration.

On their return south, Iberville opted to return to the Gulf by an alternate route that the chief had pointed out. It was a shortcut via "Bayuk Imashaka" and other waterways that would lead back to the place where they had anchored their sailing vessels.

"My brother went on in my longboat in which was the chief of the Bayogoula," Iberville reported in his March 24 journal entry. "He [the chief] gave me a Mugulasha to guide me to the sea by way of this little stream which I entered about four o'clock in the afternoon with the two bark canoes and four of my men and the Mugulasha." With great difficulty, they negotiated Bayou Manchac and then paddled down the Amite River and through Lakes Maurepas and Pontchartrain, getting back to their ships in the Gulf of Mexico two hours before Bienville's party arrived from their passage down the Mississippi.

Although the discovery of Bayou Manchac was merely a footnote to Iberville's journey, it represented the first successful European voyage along the back route. Furthermore, because of his meritorious service in exploring and claiming Louisiana, Pierre Le Moyne became the first Canadian to earn the noble French award the Cross of the Order of St. Louis.

In 1999, Iberville's adventures in Louisiana again became the stuff of legend when the state planned a commemoration of the three-hundredth anniversary of its founding as a French colony. Not surprisingly, one event in this year of recognition was an expedition to follow Iberville's path along the back route, spearheaded by Dennis Jones and Malcolm Shuman.

The archaeologists had become intrigued with Bayou Manchac and Iberville's journey while excavating the Kleinpeter site and researching area history. Theirs would be a celebratory outing, not a re-creation, to honor the anniversary, Shuman explained. "We never thought we could be Iberville." However, they planned to adhere as closely as possible to the itinerary detailed in Iberville's journal and travel as a party of five. In addition to Jones and Shuman were a fellow archaeologist, a friend with a motorboat, and a videographer who would document the historical adventure for posterity.

Much had changed in the intervening three hundred years, the team leaders knew. Most obvious was the existence and effect of the Mississippi River levee, which now blocked the bayou's junction with the river. Iberville had traveled during high water, flood season on the Mississippi, when the full length of Manchac was navigable. Nevertheless, the seventeenth-century explorer had a difficult time: "I went 2 leagues [roughly 6 to 8 miles] down it and spent the night. This river or creek is no more than 8 or 10 yards wide, being full of uprooted trees which obstruct it. . . . Within these 2 leagues I have made ten portages, some being 10 yards long, others 300 or 400 yards more or less. . . . Those portages have worn us out today, me in particular, as I had to carry the front end of my canoe and guide it for fear of having it split . . . because one of my Canadians was sick."

Equally aggravating was the disappearance of the Mugulasha guide, who simply walked away soon after they entered the stream and left the French explorers to pilot themselves along the unknown route. Pierre Le Moyne's mood and character are evident in his journal: "I have gone on nevertheless even though it is a rather venturesome undertaking with four men. . . . [I]f I turn back and go by way of the Myssysypy, I shall not catch up with my longboats and I prefer to follow this stream and show the Indians that, without a guide, I go wherever I want to go."

Unlike the Canadian hero, the tercentennial adventurers knew exactly where they were going. They were also aware that the first eight miles of the bayou from the Mississippi to the Convergence were both virtually dry and overgrown, so they would launch their expedition at the Convergence, where Iberville had discovered open water.

Jones, Shuman, and crew also eschewed seventeenth-century equipment. Instead of bark canoes, they chose a twelve-foot Coast Guard–approved aluminum motorboat and two forest-green aluminum transom canoes outfitted with small motors. Instead of sustaining themselves by fishing, hunting, or gathering, they opted for carrying large blue-and-white ice chests, a banquet of freeze-dried food, and a Coleman stove. They also hauled two gray-and-black waterproof tents, comfortable sleeping bags, a wad of detailed maps, and the many other necessities for a well-planned twentieth-century camping trip.

The party shoved off from the bank at Alligator Bayou amid the cheers of a cluster of history buffs while the smattering of local politicians who had come, sensing a photo op, beamed. The little flotilla paddled east beneath an emerald archway of fresh spring leaves. They idled along in the flat, clear water, through Manchac's stretches of wilderness beauty, and past development Iberville could never have imagined. It was a rewarding day that ended when they pitched camp, according to Dennis Jones's calculations, at the very spot on the Amite River where Iberville's party had stayed. After a pleasant night, they repacked two of their craft and headed down the Amite toward Lake Maurepas. One canoe, cursed with a faulty motor, was left at a nearby dock.

A hard east wind was blowing across Lake Maurepas, foaming whitecaps atop three-foot waves, when Jones, a seasoned outdoorsman, intrepidly paddled into the open water to survey their choices. This was the one place he was forced to make a calculated decision, as Iberville's journal hadn't specified which shore of the broad lake his canoes had followed.

As Jones perused the situation, the force of the waves smashed over the bow of his canoe, and suddenly, without warning, the canoe swamped and flipped. But in the true spirit of the twentieth century, the moment was duly captured for posterity on videotape:

one sleek, forest-green canoe hull turtled among the jerky waves of Lake Maurepas as packages of freeze-dried foods and clean clothes float toward shore. And Dennis Jones, heroic archaeologist, wet to his neck, wears an expression of frustration and disgust.

"Iberville's canoes probably were longer and heavier than ours," Malcolm Shuman surmised. And perhaps, on that March day in 1699, the winds on Lake Maurepas hadn't been quite so fierce.

Nevertheless, following their hero, who had not allowed the torturous portages or the loss of his guide to deter him, the expedition continued—only several weeks later. Beginning again at the place where the trip had been stymied, they comfortably tucked themselves inside a swift seventeen-foot speedboat that sped across the wide waters of Maurepas and Pontchartrain and arrived at Fort Pike, an early-nineteenth-century fortification at the Gulf, within a quick couple of hours.

Pierre Le Moyne himself would no doubt have been proud of their spirit.

5 The Bayou Has a Split Personality

IBERVILLE'S JOURNAL PROVIDES the earliest documentation not only of the bayou's existence but also of its peculiar hydrology. His struggle along the first eight miles and easy passage along the rest was an experience that would be duplicated many times during the next century as European colonizers sought to create the back route.

Iberville himself was the first to suggest the possibility. "It would be easy to clear out this stream during low water and make it navigable all the way to the Myssysypy," he wrote. But this strategic goal proved more formidable than anticipated, challenging monarchs and politicians, engineers and men of commerce for two centuries. And, in the end, none made the commitment necessary to overcome the natural lay of the bayou.

Bayou Manchac originated as a distributary of the Mississippi, according to geographer Fred Kniffen in his definitive 1938 study, "Bayou Manchac: A Physiographic Interpretation." After the Mississippi had migrated to its present channel, about 4,500 years ago, Manchac probably developed as the result of a break—a notch—that formed in the natural levee that banked the river. The levee and higher ground behind it had been created over the centuries as heavier sediment settled out of floodwaters. When the force of the river eventually broke through this levee, water flowed inland, tracing the southern edge of a high formation called the Pleistocene Terrace that had formed during the Ice Age (roughly 10 thousand to 1.6 million years ago), when glaciers appeared and receded. The Mississippi overflow carved this path, retaining it over centuries and creating a channel that came to join several other small waterways.

Kniffen's study revealed that Bayou Manchac had never been navigable year-round because the bed of the bayou was much higher than the bed of the Mississippi—by fifteen feet or more, depending on the season and the flood level in the great river. Only the rising tide of spring floods on the Mississippi could overcome that differential. At other times of the year, the Mississippi was too low to be a source for the bayou, so the segment of the bayou closest to the Mississippi was drained of most of its water.

Manchac also receives water from two streams that run into it from the high ground on its upriver bank—Bayou Fountain and Ward's Creek—and from the Amite River in flood stage when the flood crest flows into Bayou Manchac from the east. Because the bed of Manchac slopes down from west to east, from the Mississippi to the Amite, only the power of unusually high water can force the stream to flow backward.

Kniffen divided the channel into three segments. The first, which had challenged Iberville, ranged approximately eight miles from the Mississippi to the Convergence and historically held water only during and shortly after a Mississippi flood. The second, extending barely two miles, was from the Convergence to the junction with Ward's Creek. The third segment, the widest and most dependably navigable, was the eight miles stretching from Ward's Creek to the Amite River. Kniffen also suggested that Ward's

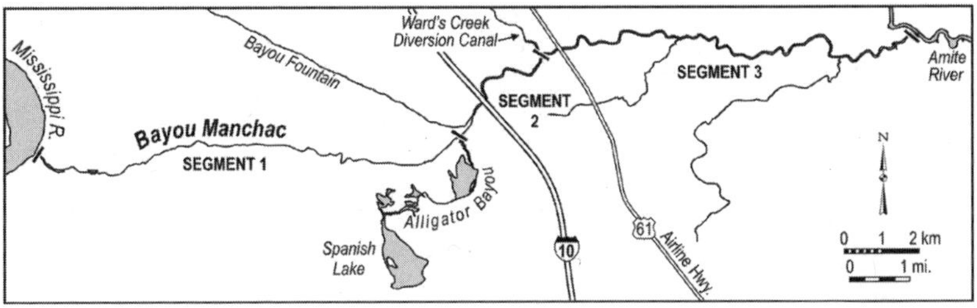

Geographer Fred Kniffen's study of Manchac divided the bayou into three segments.

Creek might once have been the primary channel, a direct outlet of the Mississippi River from the north, and older than Manchac. The newly formed Bayou Manchac, he proposed, may have joined Ward's Creek rather than the opposite.

Kniffen's original scholarship influenced how subsequent geographers would regard Bayou Manchac. But others before him, like Iberville, had already noted the peculiar hydrology. One of these was Scottish surveyor and cartographer George Gauld, who explored the then-British territory for his Royal Majesty George III. Gauld's carefully penned commentary is found in an inset on his 1769 map of West Florida: "The bed of the Ibberville is dry almost to the forks about half the year and even when the [Mississippi] river is high it is only supplied by the eddy water, but it runs with surprising rapidity within that season for the space of about 2 leagues which shows there is considerable declivity back from the river."*

British engineer and surveyor Thomas Hutchins also explored Manchac on behalf of the Crown. Though he has been accused of pirating Gauld's work, Hutchins added at least one decidedly political point: he proclaimed the Isle of Orleans a contrivance.

* The bayou's declivity is the tilt of its bed from west to east and the loss of considerable elevation from the Mississippi to the Amite.

The Isle of Orleans had been a designation first used by Bienville to claim and protect his tiny settlement of New Orleans. He declared that the "isle" extended past the surrounding swamplands to a boundary of linked bodies of water—the Gulf of Mexico, the Mississippi River, Lakes Pontchartrain and Maurepas, the Amite River, and Bayou Manchac. This placed New Orleans on an island.

The French had first used the term "Isle of Orleans" in 1754, at the outset of the French and Indian (Seven Years) War. In 1762 and 1763 when they negotiated treaties yielding their New World landholdings, the French gave the Isle of Orleans to Spain and ceded West Florida to the British. Bayou Manchac became one of the lines of demarcation between these two colonies, in effect an international boundary. In 1803, when President Thomas Jefferson

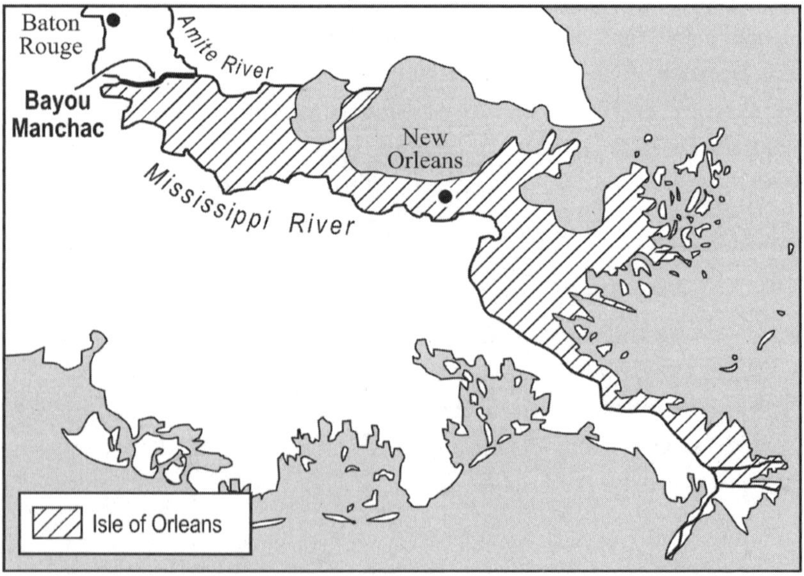

Bayou Manchac formed part of the northern boundary of the territory France identified as the Isle of Orleans.

wanted the port of New Orleans for the United States, his representatives negotiated for the Isle of Orleans.

The French had naturally disregarded the hydrology of the bayou for political reasons: there was no real island if they acknowledged that, for much of the year, Manchac was not a full stream but only a dry depression. Thomas Hutchins roundly denounced their construct:

> The city of New Orleans is not an island but on the continent. Or if the tract of land on which that city is situated can be termed an island, the name can with propriety be applied to it only two, or at most three, months every year, when the Mississippi overflows. An accidental communication with Lake Pontchartrain is then opened through the Gut of Ibberville. It may be dignified during that short period with the title of river but dries up as soon as the Mississippi ceases to overflow. At any other time, the walking from English to French-now-Spanish Manchac is perfectly dry.*

When Bayou Manchac was dammed, first in 1814 by Andrew Jackson and then permanently in 1826, it became a smaller and less navigable stream. As Lieutenant J. K. Hezlep of the Corps of Engineers described it in 1868, its first segment was "filled with logs, growing trees and undergrowth. . . . [I]t has an average width of about twenty yards but is obstructed by falling and leaning trees, logs, etc."

The following year, traveler Samuel Lockett concurred: "Bayou Manchac's upper part is growing up with cottonwoods and swamp shrubs," he wrote, but further toward the Amite "is navigable with eight feet of water for eight miles above its mouth."

The bayou today has changed little from Hezlep's, Lockett's, or Kniffen's descriptions. The sole exception is a serene lagoon

* Thomas Hutchins, *Historical Narrative and Topographical Description of Louisiana and West-Florida* (1784).

near the Mississippi, created by owners of adjacent land who have cleared a small stretch of the first segment and dammed it at their rear property line. Beyond the dam, however, for approximately seven miles to the Convergence, the channel of Bayou Manchac is dry, narrow, and overgrown, a meandering depression in the landscape where rainwater stands as an occasional murky pool. At the Convergence, however, as if by magic, the tea-colored water begins to flow toward the Amite, just as it has always done.

Bayou Manchac today
Photograph by the author

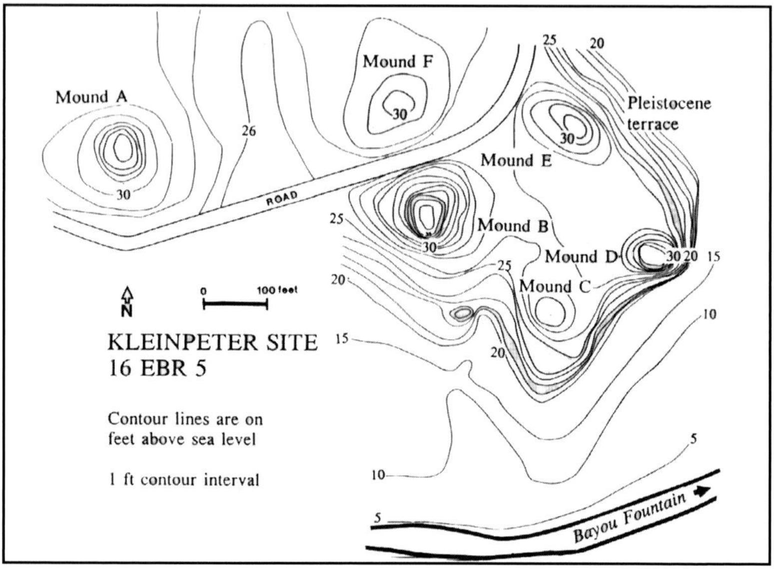

Louisiana Geological Survey diagram of the Kleinpeter archaeological site, drawn 1990.
Courtesy Louisiana Geological Survey

Mound D of the Kleinpeter site, taken from the bank of Bayou Manchac.
Courtesy LSU Museum of Natural Science

Pierre Le Moyen, Sieur d'Iberville
Courtesy State Library of Louisiana

Glenn Cambre and "the leaning oak" at the site of Galveztown
Photograph by the author

Sketch of the original San Gabriel Catholic Church as it looked in 1776, by David Broussard. The drawing is based on architectural research and physical investigation of the building.
Courtesy David Broussard

St. Gabriel Catholic Church today
Photograph courtesy David Broussard

Manchac Plantation ca. 1851, by Marie Adrien Persac

Courtesy State Library of Louisiana

Woodstock Plantation

Andrew D. Lytle Collection, Mss. 893, 1254, Louisiana and Lower Mississippi Valley Collections, LSU Libraries, Baton Rouge, La.

Dry segment of the bayou
Photograph by the author

The steamer *Alice*
Courtesy Historic New Orleans Collection

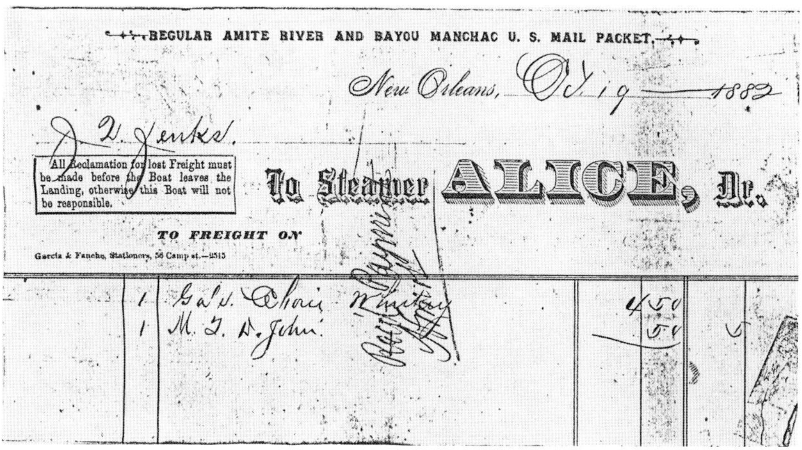

REGULAR AMITE RIVER AND BAYOU MANCHAC U. S. MAIL PACKET.

New Orleans, Oct 19 1882

J. D. Jenks

All Reclamation for lost Freight must be made before the Boat leaves the Landing, otherwise this Boat will not be responsible.

To Steamer ALICE, Dr.

TO FREIGHT ON

Receipt from the *Alice*, 1882
Livingston Parish Papers, Mss. 2898, Louisiana and Lower Mississippi Valley Collections, LSU Libraries, Baton Rouge, La.

Baptism in Bayou Manchac at Hope Villa bridge, early twentieth century
Courtesy Teri James Casso

Hand-pulled cable ferry on Bayou Manchac at Hope Villa, prior to 1917
From Baton Rouge State Times, *October 6, 1949*

Old Hickory Lodge at Hope Villa

Photograph by Elemore Morgan Sr., courtesy Elemore Morgan Jr. and State Library of Louisiana

Old-growth cypress tree, late nineteenth century
Andrew D. Lytle Collection, Mss. 893, 1254, Louisiana and Lower Mississippi Valley Collections, LSU Libraries, Baton Rouge, La.

The Caillouets' house during Tropical Storm Allison. At this time the floodwaters were receding. Earlier, the mailbox had been completely submerged.
Courtesy Pamela Stacks Caillouet

Winding through time
Photograph by the author

6 Colonial Development Begins with the French

HAVING SECURED THE LOUISIANA TERRITORY with Iberville's exploration, the French began to develop the lower Mississippi valley, concentrating settlements at New Orleans and Natchitoches and dispensing land grants along the Mississippi River. But they remained highly conscious of Bayou Manchac, which, they believed, held promise as a strategic waterway for the colony, a shortcut from the upper Mississippi valley to their settlements at Biloxi and Mobile on the Gulf.

The land at the confluence of Manchac and the Mississippi was reasonably high, and thus it was touted as a prime location for the colony's new capital. It was in fact the preferred site of John Law's Company of the West, with which the French government had contracted to develop the territory. There was even talk of channeling the bayou into a canal, perhaps inspired by the grand Canal du Midi, which had been successfully carved through the French countryside in 1681 to link the Atlantic Ocean with the Mediterranean Sea.

In April 1718, the Company of the West ordered the colonial leadership "to find the most convenient place for trading with Mobile, whether by sea or by Lake Pontchartrain . . . [with] the least danger from inundation when floods occur, and as near as possible to the best agricultural lands. These various considerations convince us, as far as we can judge, that the most convenient site is on the Manchac brook."

However, a Monsieur Perrier, the engineer who would have taken charge, was in Cuba and died en route back to Louisiana before he could implement the order. In the meantime, Bienville, now acting territorial governor, much preferred a site at a wide bend of the Mississippi substantially south of Manchac that offered its own back route. Bienville's choice, of course, became the location of the present-day French Quarter in New Orleans, which linked to Lake Pontchartrain through Bayou St. Jean (John).

As a result, the French planted no settlement at Bayou Manchac, although land grants existed nearby along the Mississippi: Diron D'Artiguette's concession located on the site of present-day Baton Rouge, fifteen miles above the bayou, and Paris Duvernay's concession that clung to the west bank of the Mississippi near the village of the Bayou Goula. Both brought new residents to the area. The French crown also enticed Tensas Indians, under siege from an alliance between the Choctaw and the British, to relocate near Bayou Manchac, in the hope that their camp might be part of the French colony's territorial defenses.

Despite the lack of a primary settlement there, the bayou became a regular route for the expansive French fur trade that ran from the upper Mississippi valley and Canada. *Coureurs de bois** and other fur traders and trappers transported their valuable cargo of beaver, otter, deer, and bear pelts down the Mississippi and through Bayou Manchac to the Gulf. The first recorded shipment contained ten thousand deer and five thousand bear hides that arrived at Manchac from the Ohio River in 1705.

* Literally "runners of the woods"—rough Frenchmen who learned from the natives how to trap and live in the woods.

Such trade was highly lucrative for Louisiana and produced an unexpected and welcome side effect: after delivering their goods, *voyageurs* sometimes stayed in the southern tier of the new French colony—welcome additions to the sparse population.

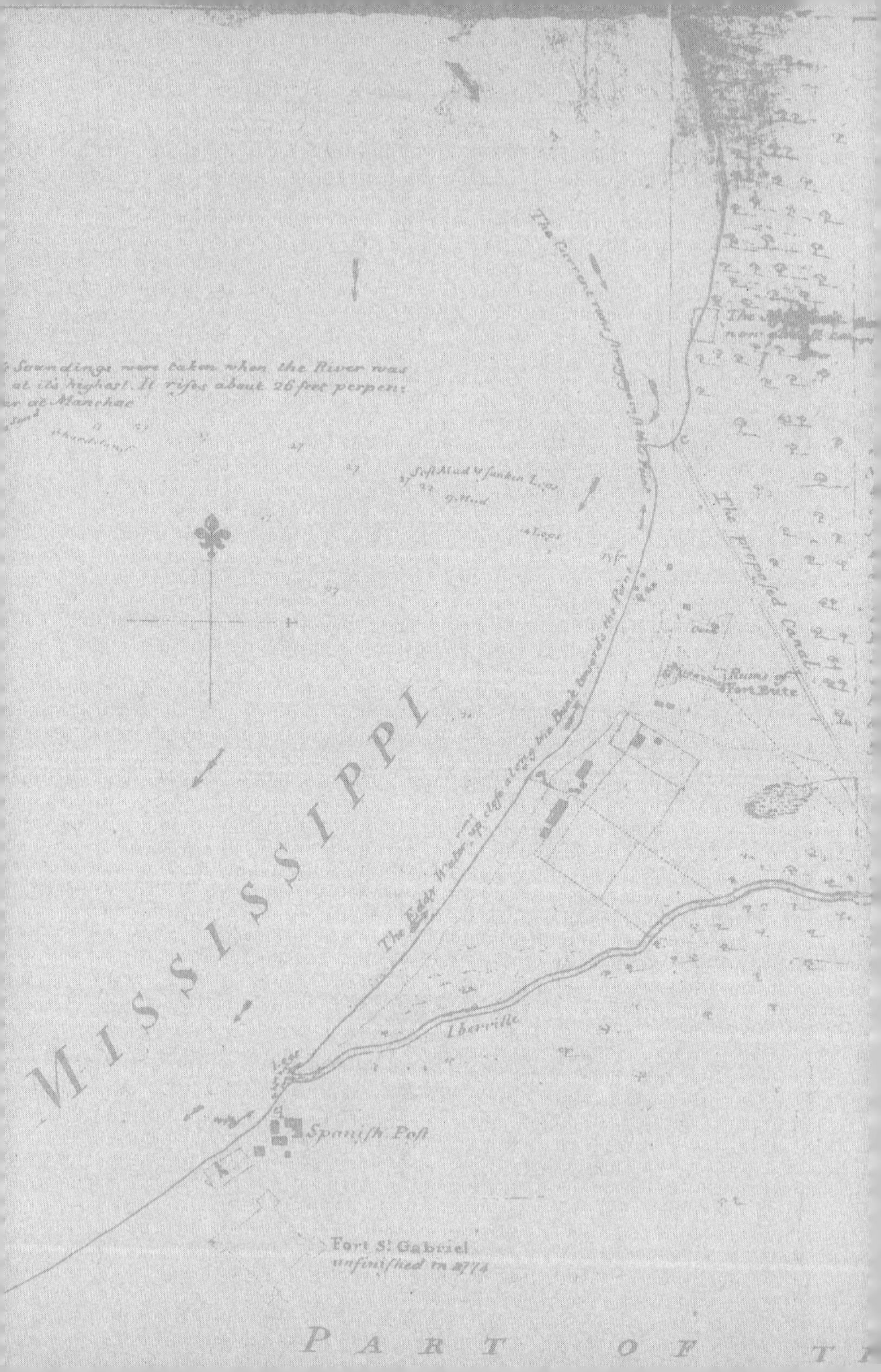

MISSISSIPPI
The proposed Canal
Ruins of Fort Bute
Iberville
Spanish Post
Fort St. Gabriel
unfinished in 1774
PART OF

7 An International Boundary

THE FRENCH CONTROLLED LOUISIANA until the mid-eighteenth century, by which time they had become embroiled on many international fronts. The colony had become a burden on their fiscal resources, so in late 1762, French monarch Louis XV secretly ceded the Isle of Orleans and the Louisiana territory west of the Mississippi to his Spanish cousin Charles III. Three months later, having lost New France (Canada) in the French and Indian War, France

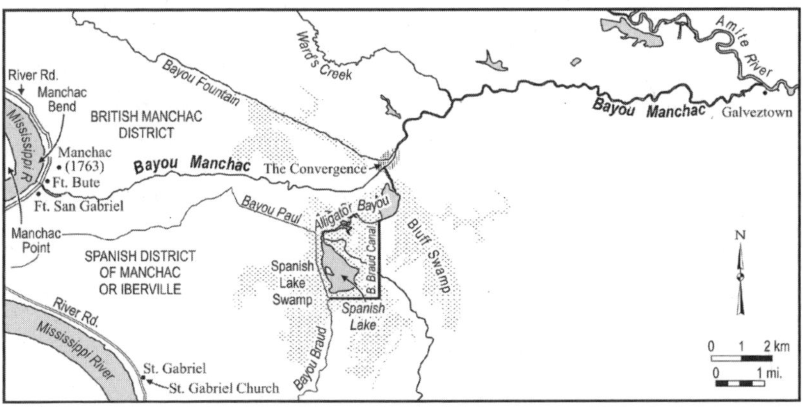

From 1763 to 1783 Bayou Manchac was an international boundary between the British Manchac District and the Spanish District of Manchac or Iberville.

relinquished its territories east of the Mississippi River to Great Britain, including (with Spain's agreement) Florida.

These transactions dramatically redrew the colonial map of North America. The narrow, winding channel of Bayou Manchac became an international border between the colonies of England and Spain; the boundary line ran exactly down its midstream.

8 The British Settle the Manchac . . .

IN 1763, THE TERRITORY OF West Florida was one of Britain's three new American acquisitions, along with East Florida and Quebec. West Florida extended from the Appalachicola River on the east to the Mississippi River on the west, from the Gulf of Mexico on the south (except for the Isle of Orleans, of course) to the line of 31 degrees north latitude on the north.* Pensacola, Mobile, Baton Rouge, and Natchez were designated as its capitals.

THE BACK ROUTE AND FORT BUTE

The Treaty of Paris allowed British trade on the Mississippi River. But because the Spanish controlled the port of New Orleans, the British believed that access to the Gulf through the Iberville River and the back route, bypassing New Orleans, offered a strategic and more dependable alternative. Opening the back route, Great Britain hoped, would cut off Spain's access to trade from the upper river valley as well as shorten the distance for British traders from upriver to Mobile, Pensacola, and Caribbean ports.

* The latter, an arbitrary boundary, was edged further north to the mouth of the Yazoo River in 1764.

In October 1763, an initial attempt was made to open the first segment. Captain James Campbell and a small contingent of soldiers along with fifty slaves worked tirelessly to clear the streambed, cutting back the thick cane that grew along the banks, chopping fallen trees that lay across the channel, and removing logs and driftwood swept into the bayou during the last Mississippi River flood. By December, Campbell reported to his superiors that the Iberville River had been made navigable from end to end.

The British were clearly unfamiliar with the bayou's annual cycle, however, and Campbell's accomplishment was short-lived. When annual floods arrived as usual in March, trees and driftwood were once again swept into the bayou and piled up against debris that had been left farther along the channel. This created a logjam even greater than the one they had cleared.

In the fall of 1764, the new provincial governor, British navy captain George Johnstone, championed the Iberville River as "the hub of [West Florida's] commercial glory" and advocated construction of a fort and habitations at the junction of the bayou and the Mississippi. "There is no place of so much consequence to this province as that settlement now the Iberville is open'd," he enthused. "[It] will command the whole trade of the Mississippi."

In 1765, Lieutenant Philip Pittman of the 34th Regiment of Engineers was dispatched to evaluate the potential for Mississippi River development. His report supported Johnstone's assessment and noted that both Indians (promising trading partners) and other residents along the river and its tributaries "would rather trade [at Manchac] than at New Orleans if they could have as good returns . . . for it makes a difference of ten days in their voyage."

As construction of a fort began in February 1765, Johnstone reiterated his conviction about the necessity of clearing the first segment to make a navigable back route. "The further prosperity of the province depends upon it," he wrote, especially in light of the delayed takeover of the Isle of Orleans by the Spanish. This had left Britain's enemy, France, in power long past the signing of the treaty.*

Johnstone tactfully recommended that the six-gun, star-shaped earthwork post should be named Fort Bute, to honor the British prime minister, John Stuart, Lord Bute. The small, wooden, stockaded fort with a blockhouse would "accommodate a captain, fifty men, and appropriate artillery," with the capacity to house an additional two hundred men, a worthwhile precaution. Johnstone also ordered two ships docked in the Mississippi River near the fort for additional protection.

The first assault on the fort came in August 1765 when Alabama Indians who lived in a nearby village (perhaps at the same location to which Iberville's Mugulasha guide had repaired when he abandoned them) attacked the British force, stole whatever supplies and guns they could, and killed livestock. All the occupants of the fort fled.

But Johnstone was deeply committed. He ordered Fort Bute rebuilt and dispatched more soldiers. Still, according to Captain Harry Gordon, who arrived at Manchac in October 1766, it remained a sad little installation:

> [The fort] is a Square with half Bastions—they had better been whole ones—of bad Stockades. There are Huts in the inside for Officers and Men, 100 in Number. The Intention of this Post is to cover our Communication to the Mississippi by Pontchartrain and Maurepas Lakes and thro' the Gully or Ditch of the Ibbeville, where there is any Water in it which is only the case when the Floods come down the river. The Bed of this [the Mississippi]

* Though the Spanish were ceded the Isle of Orleans in 1763, the first Spanish governor would not arrive in New Orleans to assume control of the colony until 1766.

> was now 24 Feet under that of the Ibbeville. We Endeavored to View the clearing of [the Iberville] this last but were only able to go along it for 3 Miles on account of the Rankness of the Weeds, thro' which there is no Path.

By 1768, Fort Bute had once again been abandoned.

BRITISH DEVELOPMENT—AND A CANAL?

To protect and defend the colony, the British needed to attract additional settlers. One strong enticement was the offer of free land grants. In addition to increments of five thousand acres or more that were tendered to veterans of the French and Indian War, one hundred acres could be claimed by any head of a household, with an additional fifty acres for other household members, and a thousand acres could be had by anyone who promised to put the granted land into cultivation, especially in indigo.

Despite impenetrable forests and swampland teeming with mosquitoes and alligators, the rich land around Bayou Manchac was deemed more appealing for agriculture than the sandy terrain around Mobile and Pensacola. And, since both military and commercial interests continued their campaign to make the Iberville fully navigable, the land held promise of becoming more valuable than land along just any backwater meander. So while land grants did attract some settlers, they also appealed to investors who just wanted acreage with an eye for profit but who probably never intended to live on or develop their tracts.

Lieutenant Pittman's report* became a tool for helping promote West Florida. In it he provided a detailed survey of the back route for his superiors, most of whom had never set foot in the

* *The Present State of European Settlements on the Mississippi*, published in 1770.

territory. Pittman had entered Bayou Manchac from the Amite River, finding it "necessary to strike the masts when the waters are high as the branches of the trees hang very low and in some parts they form an arbour over the river." His report continued:

> At irregular distances, as from one to three hundred yards on each side of the river, there are high lands overgrown with canes. . . . one league higher is Anatamaha [the Convergence] which in the Indian language signifies the fish place: it is properly called for they abound here all the year which accounts for the vast numbers of crocodiles that are continually on the banks of this river. Sloops and schooners may come as far as this place when the waters are at the lowest . . . never less than twelve feet [of] water. . . .
>
> From hence to the Missisippi I think the trees should be cut down forty feet back from the river side that a road might be made for carriages when the waters are low. . . . When the waters are high it will be necessary for the navigation as vessels may be tracked up by horses or men to the Mississippi in the same manner as lighters in England.* My bateau went about one mile above Anatamaha but I could not get her any higher, although there was not less than four and five feet [of] water all the way to the Mississippi which is more than three leagues distance . . . and when the waters are up there will be from eleven to eighteen feet [of] water.

The first segment clearly remained an obstacle course.

Pittman's solution, like that of the French before him, was a canal. He would have known of the Canal du Midi as well as Britain's own Bridgewater Canal, begun in 1759 across northern England, and he realized that a canal might effectively open Manchac and create a true back route. He offered concrete evidence for his

* A lighter, a type of flat-bottomed barge, was traditionally steered with oars, but Pittman envisioned their being pulled, as they apparently were in contemporary England, because the bayou would be too deep for poling and too narrow for rowing.

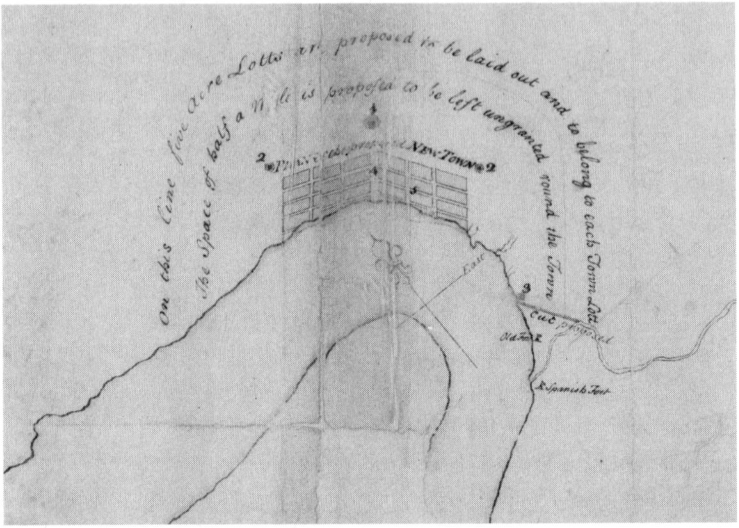

Detail from Elias Durnford's 1771 map that includes the proposed town of Harwich. The town is laid out at a bend in the Mississippi River. The small stream at the far right is Bayou Manchac (designated on another portion of the map "River Ibberville"). Also in the bottom right quadrant, the words "cut proposed" mark the planned canal between the river and the bayou. *Courtesy of LSU Libraries' Special Collections*

argument by measuring distances from Manchac to British settlements on the Gulf. Snaking down the Mississippi River to Balize at the mouth of the Mississippi was sixty leagues; sailing the back route all the way to Pensacola was only ninety leagues.*

Pittman recommended cutting a true canal on an angle from the top of Manchac Bend above the bayou's junction with the Mississippi to a place farther east along the first segment. He predicted that this would take advantage of the river's swift current, filling the canal and avoiding the problem created at the natural junction on the outside of the bend, where a swirling effect during flood season sucked in logs and driftwood.

* A league was a measurement that could be any length from 2.4 to 4.6 miles.

This seemingly sensible solution offered real promise for transforming the River Iberville and required only political commitment and money from the British government. They accepted Pittman's report and proceeded to debate the issue from 1771 until the outbreak of the Revolutionary War.

The British Board of Trade, reflecting the interests of merchants and other well-connected citizens, was in full support of Pittman's plan. They also favored reestablishing Fort Bute (to keep a watchful eye on the Spanish) and creating a more formal settlement at the junction of the bayou and the Mississippi to enable Britain to take better advantage of the lucrative fur trade. Some £100,000 sterling in furs was received annually at the Spanish customhouse in New Orleans, trade the British hoped to win.

Meanwhile, political opponents of Iberville River improvements, including surveyor George Gauld, were equally vocal. Among their arguments: the British military command had withdrawn its troops from the Illinois country, so the back route was no longer needed to facilitate troop movement from the Gulf to the interior.

As the debate continued, the new colonial governor, Peter Chester, began planning the new town near the possible canal. The settlement would be named Harwich and serve as a warehousing and trading center. Dartmouth, a sister settlement, was also proposed for construction at the other end of Manchac, at its junction with the Amite.

Publicizing the plan of Harwich led to the purchase of eighty-two town lots. Terms of agreement included building a house with a brick chimney within two years or facing a penalty. But again, interest fell short of expectations and many buyers were speculators. Even British refugees from the pre-revolutionary turmoil on the Atlantic coast who were considered a potential market for Harwich lots failed to participate. (The majority of them chose instead to take up residence in the province of East Florida.) As a result, neither Harwich nor Dartmouth ever developed beyond notes on paper.

JOHN FITZPATRICK, MERCHANT OF MANCHAC

Despite the spasmodic nature of area development, however, a tiny community known as Manchac had emerged near the fort. Its residents included several successful merchants who had built a thriving trade with Indians from the upper valley, with British colonists on the Atlantic seaboard, and (illegally) with Spanish colonists on the Isle of Orleans.

Among the most successful was John Fitzpatrick, an Irish Catholic who had first immigrated to the Illinois country upriver, then to French New Orleans, and finally to Manchac, after Spanish governor Alejandro O'Reilly expelled British nationals (among others) in 1768. Fitzpatrick owned a store at Manchac with a house attached and had established a warehouse at the Convergence to facilitate the transfer of goods. He was also the proprietor of an indigo and tobacco plantation just upriver from his store.

Fitzpatrick traded in a broad range of goods—hides and furs, tobacco, tallow, bear's oil, sugar, flour, corn and other foodstuffs, hardware, liquor, household effects, gunpowder, blacksmith tools and more. He bought and sold widely and was reputed to have among his customers the pirate Jean Lafitte, who was said to have bought beaver skins and tobacco from him.

Fitzpatrick and the several other British merchants in the area were well received by residents of the Isle of Orleans, who were happy to have manufactured goods and foodstuffs readily available, regardless of their source. Early Spanish governors pretended not to notice the trade. In fact, trade "by way of Little Manchac" became late-eighteenth-century shorthand in the Spanish colony for transactions not entirely licit. But when Bernardo de Galvez assumed control of the colony in 1777, he declared English commerce at New Orleans detrimental to Spanish merchants and expelled any English entrepreneurs and their representatives still operating on the Isle of Orleans.

Despite the merchants' success, however, Manchac was a struggling community. When Lieutenant Rufus Putnam visited in April 1773, he observed: "We Landed [at Manchac], found it a Small crik with So little Warter at this season that many thousand fish ware Dead and made a Very Disagreeable Smell. [On] the Spanish side was an officer and 10 men. But none on the English. [There] was one Very good House on the English Side and Some tolairable gardens." Throughout his journal, however, Putnam refers to sloops and men coming and going from "Manshack" with no further explanation of the place, which seems to indicate that both the Iberville River and the settlement of Manchac were familiar places to those who would read his communications.

BARTRAM'S EXPEDITION

Perhaps the most renowned visitor to British Manchac was botanist William Bartram, who arrived two years after Putnam and reported much more kindly about the outpost. Bartram, privately contracted by London physician and collector John Fothergill, was traveling on a four-year commission and would be the first to collect and draw the native flora of the southeastern British colonies.

In October 1775, Bartram traveled the back route by boat from Mobile and entered Bayou Manchac from the Amite River, landing at the Convergence. "We soon came to the landing where are warehouses for disposing merchandize, this being the extremity of navigation of this canal and here small vessels load and unload," he wrote in his *Travels*. "From this place to Manchac, on the banks of the Mississippi . . . is nine miles by land. The road straight, spacious and perfectly level under the shadow of a grand forest; the trees of the first order in magnitude and beauty."

This road was the carriage trail Pittman had proposed, built along the north bank of the bayou from the Convergence to the Mississippi and used to transport goods and people when the first segment of the bayou was too low for shipping. Though inefficient, the overland link connected boats from the lakes and the Amite River and boats that plied the Mississippi. Bartram, however, makes no mention of whether he covered the ground on foot, by horseback, or in a carriage.

Whichever means he used to pass through the "grand forest," however, the exuberant naturalist recorded detailed observations of the large and beautiful trees along the way, using their Linnaean names, omitting their common ones. His list included *Magnolia grandiflora* (southern magnolia); *Liriodendren tulipfera* (tulip tree); *Plantanus occidentalis* (sycamore); *Juglans nigra* (black walnut); *Fraxinus excelsior* (white ash); *Morus rubra* (red mulberry); *Laurus borbonia* (red bay); *Telea basswood* (wafer ash); *Liquidamber styraciflua* (sweet gum), and others.

Indeed, "trees of the first order in magnitude and beauty" still grow along the less-developed banks of the bayou, I learned while on an outing with LSU botanist Lowell Urbatsch. Although he had botanized much of Louisiana, Urbatsch confessed with an embarrassed grin that he had never seen Bayou Manchac before.

We puttered along the stream in a little motorboat as Urbatsch called out the common names of the trees he could identify along the banks: "American elm . . . green ash . . . honey locust . . . green hawthorn . . . hackberry . . . sugarberry . . . possum haw . . . red maple . . . water oak . . . basket oak . . . live oak . . . cypress . . . cherrybark oak . . . cedar elm . . . red mulberry. I had no idea all this was out here," he marveled. His observations continued unabated for several miles until, unfortunately, the motor became clogged with duckweed that was as thick as carpeting on the water, which brought our adventure to an abrupt end.

William Bartram, however, met no such fate. He reached the village of Manchac on October 21, 1775, and he took his first sighting of the Mississippi River:

> I stood for a time, fascinated by the magnificence of the great sire of rivers. The banks of the river at Manchac, though frequently overflowed by the vernal inundation, are about 50' perpendicular height above the surface of the water. These cliffs, the natural accumulation of sediment from annual flooding, were a light, loamy consistency, continually cracking and parting. . . . There is yet visible some remains of a high artificial bank [a levee] in front of the buildings of the town. . . . The Iberville in the summer season is dry and its bed is 12 or 15 feet above the surface of the Miss. But in the winter and spring has a great depth of water and a very rapid stream which flows into the Amete.*

The naturalist noted a few impressive buildings at Manchac, including the warehouses of Swanson and Company, one of John Fitzpatrick's rival merchants. And he took careful note of Britain's neighbors across the Manchac channel: "The Spaniards have a small fortress and garrison on the point of land below the Iberville, close by the banks of the river. The two banks are connected by a slender narrow wooden bridge across the channel supported on wooden pillars and not a bow shot from the habitations of Manchac."

From Manchac, Bartram proceeded north along an overland trail that followed the natural high ground near the Mississippi.

* In 1931, graduate student Edith Singletary climbed the old levee while doing thesis research, seeking the place where the bayou met the Mississippi. There were no cliffs remaining such as Bartram had noted, but she discovered "a depression filled up with willows and gum trees, where an old ditch once was." That ditch was no doubt the relic channel.

Today, however, this evidence is gone. Because eastward movement of the river in this area has compelled several levee setbacks, everything that Bartram saw is now approximately a half mile into the Mississippi River.

He arrived at an "Alibama Indian village" about two miles above Manchac, "delightfully situated on several swelling green hills, gradually ascending from [the riverbank]." His itinerary continued upriver, into what is now the state of Mississippi, before he retraced his route and returned to Manchac. There he deposited his collection of "growing roots, seeds and curious specimens" with the merchant Swanson to send back to Fothergill in London.

YEARS OF THE AMERICAN REVOLUTION AND ITS AFTERMATH

Some historians have suggested that William Bartram made his journey through this new British territory to avoid the growing unrest in the thirteen Atlantic colonies. If so, he would have found British West Florida a welcoming place, for its settlers evidenced little public involvement in the revolutionary activities of their fellow colonists against the crown.

West Florida was not entirely spared the fervor of the American Revolution, however. In 1778, Manchac received another visitor, one deeply engaged with the cause. Captain James Willing was a Natchez plantation owner and merchant sympathetic to the Revolution. With a small force of political cohorts, Willing sailed the Mississippi River in his barge, the *Rattletrap,* burning plantations owned by British colonists, capturing their slaves, and pursuing British ships along the Mississippi. Rumors of Willing's activity so terrified some British plantation owners that they fled across the Iberville River into Spanish territory.* Merchant John Fitzpatrick was among them.

* William Dunbar, a prominent planter who lived along the Mississippi above Bayou Manchac, wrote a vivid description of Willing's raid, of which he was a victim: "All was fish that came into their nett. [They seized] all my waring apparel, bed and table linen . . . not a shirt was left in the house." The raiders also stole "blankets, pieces of cloth, sugar, silver ware."

The outbreak of the Revolutionary War consumed the British both militarily and financially and forced them to focus on their Atlantic coast colonies. But Willing's raid along the Mississippi exposed a weakened flank; it was apparent that Great Britain needed to turn some attention to West Florida. In March, a small detachment of soldiers appeared at Manchac to refortify the post.

The American Revolution changed geopolitics in both the New World and Europe. The French crown, Great Britain's historical rival, supported the American colonies and was likewise allied with Spain. This meant that the Spanish on the Isle of Orleans, just across Bayou Manchac, were Britain's avowed enemy, which terminated any fraternizing with the residents across the streambed "just a bow shot" away.

Although the British had never transformed the first segment of Bayou Manchac into a navigable channel or built the canal that sounded so promising, His Majesty's government nevertheless sent General John Campbell to strengthen Manchac. His mission was to enlarge the post to accommodate three hundred soldiers as part of a secret British plan to seize the Isle of Orleans and Louisiana territory from Spain.

But in June 1779, before England could declare war on Spain, Spain declared war on England. The British in West Florida, however, did not learn of it until September, and by then, Spanish governor Galvez was marching north along the Mississippi from New Orleans, headed for British Manchac. He had learned that the British were strengthening their forts at Manchac and Natchez and had rightly assumed their motive was to launch a surprise attack rather than simply to defend West Florida. So he wisely chose to go on the offensive.

As Galvez marched upriver, his force was joined by Acadians

and Germans from the settlements along the Mississippi, amassing as many as a thousand men. Alexander Dickson, the British colonel in charge of the Manchac District, realized he was greatly outnumbered and could not defend the Manchac post against Galvez's army. Dickson moved most of his force to the larger Fort New Richmond at Baton Rouge so that, when Galvez and his polyglot militia arrived at Manchac, they overwhelmed it with little bloodshed. One British soldier was killed, eighteen taken prisoner, and the Spanish army rested for a few days before continuing their march upriver.

The siege of Fort New Richmond took less than a day, and shortly thereafter, Galvez's troops captured the British post at Natchez. Other British posts above Baton Rouge surrendered as well, and Galvez returned to New Orleans, completing plans to capture the strongholds at Mobile and Pensacola. Indeed, by mid-May 1781, he had successfully brought West Florida back under the Spanish flag.

The British residents of Manchac who chose to stay took an oath of loyalty to the Spanish king as the population of the little community grew—from 77 residents in 1785 to 284 residents in 1788. Many of these were farmers, who could now legally sell their produce in New Orleans as there was no longer an international boundary to cross.

It is tantalizing today to regard the sleepy landscape where the bayou meets the Mississippi and wonder what might have happened if the British had pursued Philip Pittman's seemingly practical plan for a canal. Instead of pastureland, might a thriving commercial settlement anchor an important junction on a busy back route to the Gulf? Would Bayou Manchac subsequently have been straightened, widened, and dredged for large vessels, its banks now lined with industry? Would it be today a visible and well-known stream?

History revolves around such questions.

9 And the Spanish Settle the Manchac

ALTHOUGH THE BRITISH had assumed control of West Florida immediately after receiving it from France in 1763, the Spanish were slow to take possession of the Isle of Orleans. This delay occurred in part because they lacked sufficient troops to assign to the new colony, but during the interim, the French population of New Orleans grew hopeful that Spanish rule might never need come to the Isle of Orleans. The appointment of Charles-Philippe Aubry as acting governor and agent for the Spanish only fanned their optimism.

SAN GABRIEL, THE FIRST SPANISH SETTLEMENT

This situation inevitably made matters difficult for Antonio de Ulloa when he arrived in March 1766 to assume his duties as governor. Ulloa moved deliberately, retaining Aubry while he took time to examine his new colony's status, which he discovered was in dreadful condition. Most buildings needed repair and few defensive posts existed to protect Spanish interests, including at Bayou Manchac just across the channel from the British Fort Bute.

Ulloa dispatched soldiers to build a modest four-gun stockade at Manchac since, unlike the British, he had no need to utilize the bayou for trade. He did, however, need to attract settlers to populate the area.

The first arrived in July 1767. Fifty Acadian families (French citizens expelled from Nova Scotia) were sent to Fort San Gabriel de Manchac. Recently attracted to New Orleans from Maryland, the 210 men, women, and children were sent pioneering by the Spanish government, given land grants, food rations, tools, seeds, and ammunition ("one gun, twelve gun flints, three pounds of powder, and bayonets" for each family). In exchange, they would confront the English across the bayou, fend off Indian raids as necessary, and cope with mosquitoes, malaria, and flooding.

The community grew slightly in the suceeding months. By the time the fort was dedicated in January 1768, two companies of soldiers had joined the Acadians and, during the following year, a group of German immigrants arrived to settle there as well.

Although Ulloa initiated policies to strengthen the colony's economy, the Creoles of New Orleans grew increasingly resentful. In 1768, they rebelled and, having few troops to protect himself, the governor was forced to flee the Louisiana territory. Such was the ignominious beginning of Spanish rule.

Aubry was appointed once again to serve as the colony's leader until Ulloa's replacement, Lieutenant General Don Alejandro O'Reilly, arrived. O'Reilly, an Irish-born Spanish resident with long service in the Spanish army, made his appearance in New Orleans in 1769 with an impressive show of force. He brought with him two thousand soldiers, an array of artillery, and a determination to bring the rebellious colony into line.

The new governor gave amnesty to most of the known rebels who had forced Ulloa's ouster but had several of the ringleaders tried and executed. He implemented other firm measures meant to establish his authority and ordered all inhabitants of Spanish Louisiana to take an oath of allegiance to Spain or leave the province.

When O'Reilly departed, the Spanish crown was firmly in control of its Louisiana territory.

Luis de Unzaga, O'Reilly's successor, decided that Spain required a greater show of strength at Manchac. An appraisal of the fort in 1774 had reported it "worse than 1770," merely a stockade and a small collection of useless cannons. So in 1775, Unzaga ordered the fort rebuilt and strengthened. He ordered a site thirty-six feet from the bluff of the Mississippi River, protected by "a rampart of cypress stakes and semi-bulwarks," to include a house, kitchen, latrine, and henhouse for the commander; a kitchen and latrine for the troops; an ammunition storage area; and a chapel.

The stability of the tiny Spanish community at Manchac depended not only on the fort but also on retaining the settlers. Governor O'Reilly had promised them a church of their own—a strategic decision to encourage the Catholic populace to feel allegiance to their parish and to the Spanish province.

The church parish of San Gabriel extended from the fort at Bayou Manchac down the Mississippi River to the boundary of Ascension, an older church parish. The San Gabriel church itself would not be located hard by the fort but on natural high ground slightly downriver where the government laid out eighty-five land grants. By 1772, residents had framed up twenty-seven houses and used a chapel in the presbytere (the priest's residence) for religious services until they could find qualified builders for a real church.

Local legend long maintained that the San Gabriel Church had originally been built next to the fort on the bayou and was later removed to its present location several miles downriver. But contemporary historians have debunked this. They believe the confusion arose because the church was actually moved back on its own property several times. This was a common practice along the lower Mississippi River. When a new levee was constructed to accommodate changes in the course of the river, existing buildings would be set on logs and rolled back on their own property.

The residents of Spanish Manchac were taxed 1½ piastres for each front arpent of land they held to help build the church, and they also helped with construction.* Under the leadership of Louis LeConte, the selected contractor, the church of San Gabriel de Manchac was completed in August 1776. The simple Creole-style building of native cypress consisted of a small interior sanctuary, 60 feet by 33½ feet, surrounded on all sides by a dirt-floored exterior gallery. "It is very beautiful," wrote Commandant Louis Dutisne to the governor, an opinion shared by Fortescu Cumings, who journeyed along the Mississippi in 1807. "A handsome Roman chapel called Manshack, about thirty leagues above New Orleans," Cumings wrote of the landmark.

The church of San Gabriel at Manchac is thought to be the oldest extant church edifice in the Mississippi valley and is the only surviving colonial building in the Manchac area. It has been much altered by the loss of its exterior gallery and the addition of a New England–style steeple, but the essential framework of the old church remains. When a new St. Gabriel Catholic Church was built in 1953, the original bell, cast in Spain in 1771, was moved to the new belfry, where its inscription can still be read: "Sancta Maria de la Merce ora pro nobis; Se hizo siendo comendador el M.R.P.I. Ygnacio de Jesus Maria, Ario de 1768." (Holy Mary of Mercy pray for us; This bell was commissioned to be made by the M.R.P.I. Ignatius of Jesus Maria, in the year 1768.)

* An arpent was the standard measure of land division used in Louisiana along navigable waterways. It gave every property owner some frontage on the water; land value was determined by how broad the water frontage was. A lineal arpent was approximately 192 feet. Each property was long and narrow; the standard depth was forty arpents back from the waterway or sometimes a double depth of eighty arpents.

According to the 1795 census, the population of the District of Manchac, the political entity corresponding to the parish of San Gabriel Church, was 304 individuals—222 whites, 1 free black, and 81 slaves. The census also noted that they farmed indigo, rice, corn, and beans. But perhaps most revealing is the list of residents' names, reflecting the strikingly multicultural heritage that was evolving. Juan Baptista Heber, Pedro Labauve, Juan Templett, Franco Daigre, and Juan Baptista Klempeter (perhaps a misidentification of Kleinpeter), among others, gave certain evidence that Acadians, Germans, and Spanish were together creating the backbone of Spanish Manchac.

The community itself has survived, though with several name changes. When the first post office was established in 1813, it was known as Iberville; in 1857 it was renamed St. Gabriel, as it is still known. The small, scattered settlement, has been incorporated and is clustered with the nearby towns of Sunshine and Carville. The Acadian houses built near the church have long disappeared, replaced by a line of petrochemical plants that loom above the little wooden church that has survived since 1776.

GALVEZTOWN

As Spanish Manchac developed, King Carlos of Spain received a detailed memoir from Don Francisco Bouligny, a military and political leader in colonial Louisiana. Bouligny had returned to Spain in 1775 for a year-long visit and brought many strong opinions about the province to his royal majesty. He presented them in a report called "Memoria," which included his belief that defense of Louisiana was grossly deficient. "The English can reach the Mississippi at Manchac through the lakes without the knowledge of the Spaniards [who] have no establishments on the banks of the lakes," he wrote. "We should [also] rebuild the fort that we have at Spanish Manchac which is distant only a pistol-shot from the English Manchac."

Bouligny was a great booster of the Spanish colony and encouraged the crown to invest more robustly in its well-being. "It is almost impossible to conceive the abundance of the forests," he wrote. "From Manchac to the seashore, only in the island of Orleans, there are 100 square miles of cypress trees as thick as the hair on the head. . . . There are few countries in the world where so many live oaks could be found, a kind of wood that is superior to any other for the interior construction of ships."

He warned the king about the English traders, who "have continually in the Mississippi ten or twelve boats, one or two floating stores, and furnish the planters with what they need. Articles are sent to [British] Manchac to form there a depository from which they furnish goods to others of our people." He also lamented the dearth of native tribes friendly to the Spanish in comparison with the number of tribes who had befriended the British.

Governor Bernardo de Galvez subsequently addressed the problem of the British traders by expelling them from the colony. But as Bouligny noted, the British could still sail unimpeded along the back route (or as unimpeded as an unimproved Bayou Manchac would allow). In November 1778, in response to Bouligny's concern, Galvez undertook a reconnaissance: after visiting the fort at San Gabriel, he traveled east to the Amite River to better comprehend the geography of his province and ascertain its weaknesses.

At the junction of Bayou Manchac and the Amite River, the governor discovered a tiny community of Anglo-Americans who had fled Canewood, a British settlement located nearby on the Amite. They were living as squatters on the bit of high land at the junction of the two waterways, a place that George Gauld had

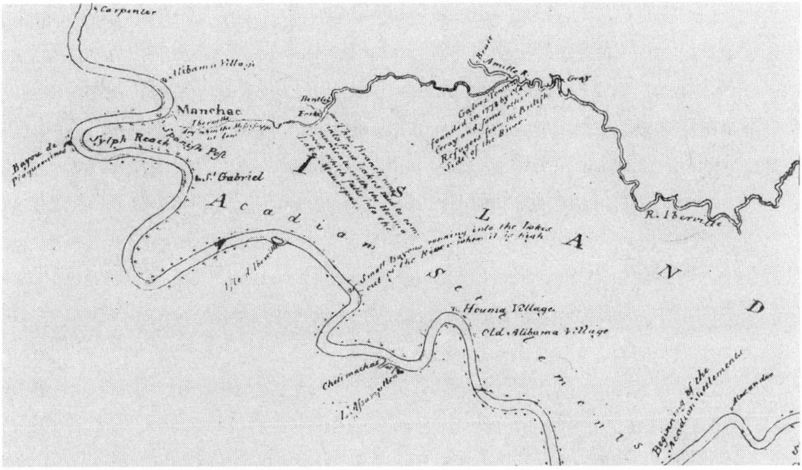

Detail of George Gauld's 1779 map of West Florida, showing the settlements Manchac and Galveztown on the Iberville River. Under the segment of the Iberville near the Mississippi, Gauld wrote "dry when the Mississippi is low." *Courtesy of LSU Libraries' Special Collections*

noted on a map of West Florida as "Galvez Town founded in 1778 by Old Gray and some other refugees from the British side of the river."

Galvez determined that the site offered a suitable location for a fort from which his military could control the mouth of the bayou and counter British trade on the Amite. He allowed—perhaps encouraged—the squatters to remain and planned to send his own countrymen to join them, to establish a community for the defense of Spanish Louisiana. In the hopes of ingratiating themselves to the governor, or in recognition of his visit, the handful of inhabitants named their location Villa de Galvez—Galveztown.

Bouligny, now lieutenant governor, attempted to dissuade Galvez from settling the site with Spanish colonists. He feared the swamps and potential floods and predicted extraordinary hardships. But Galvez prevailed.

Attorney Glenn Cambre represents the fourth generation of his family to live at the bend of Louisiana Highway 42 where it curves south along the junction of Manchac and the Amite River. A historical marker stands along the narrow road in front of Cambre's property, noting the general area where the Spanish fort and village of Galveztown once stood.

Although the buildings are long gone and the town grid has disappeared, Cambre somehow feels its presence within his modest compound, a couple of acres dotted with several white frame and cinder-block buildings. His pride in this historic site is backed by a modest trove of artifacts dug from his front yard: Hessian belt buckles, eighteenth-century square-headed nails, gunmetal cuff buttons, and numerous misshapen minié balls, all relics of Spanish occupation.

As Cambre tells the story, Spanish settlers asked native Indians for the best location to place their fort. Or perhaps it was the British refugees who had already done so—he isn't sure. But the place was identified as "at the leaning oak," the highest point in the area.

The leaning oak is in Cambre's backyard. It is a massive, split-trunked live oak with two huge limbs growing almost parallel to the ground at a height of approximately eight feet, creating the illusion of a low entrance to an unseen chamber. The tree is surely ancient and would have many tales to tell, including, perhaps, the same one that Glenn Cambre relates.

Through the picture windows in the Cambres' living room is a panoramic view of the broad, tree-lined Amite River. But from the bottom of the steep hill that contours his backyard, a glance to the left past a thick grove of trees brings clear sight of the junction of Bayou Manchac and the Amite.

"We're the highest property around," Cambre says with a gesture toward neighbors on both sides. "We've never flooded even though fifty yards that way and a hundred yards the other way often go under." His tiny island of security, however, is too small to have contained all of Galveztown—its six-street-square grid, the fort, and the acres of cultivated fields that surrounded the village. And flooding was among the principal causes for the demise of the village.

In January 1779, the first Spanish settlers sailed from New Orleans through the lakes to land at the junction of the Amite River and Bayou Manchac. The group consisted of soldiers and fourteen families from the Canary Islands, who became known in south Louisiana as Isleños.

The Canaries were a Spanish colony located on a beautiful volcanic archipelago approximately one hundred miles off the coast of Africa. Its economy was in dire straits in the mid-eighteenth century just as the Louisiana colony was in need of additional population, so the Spanish crown offered qualified men the opportunity to relocate. Criteria for selection included age (between seventeen and thirty-six years old) and appearance—at least five feet one-half inch tall, robust, and "without noticeable imperfections."

The Isleños journeyed to Galveztown with promises of land, food, farm implements, and supplies in exchange for their transforming the wilderness campsite into a Spanish outpost. They would also undertake farming in terrain and climate that was vastly different from their native land. And to connect with other parts of the colony, they could sail to New Orleans through the Amite River and the lakes or travel to San Gabriel by a primitive fifteen-mile overland trail.

The village of Galveztown was laid out under the direction of Commandant Francisco Collell on a classical plan—lots plotted around a square plaza. On each lot would be built an identical

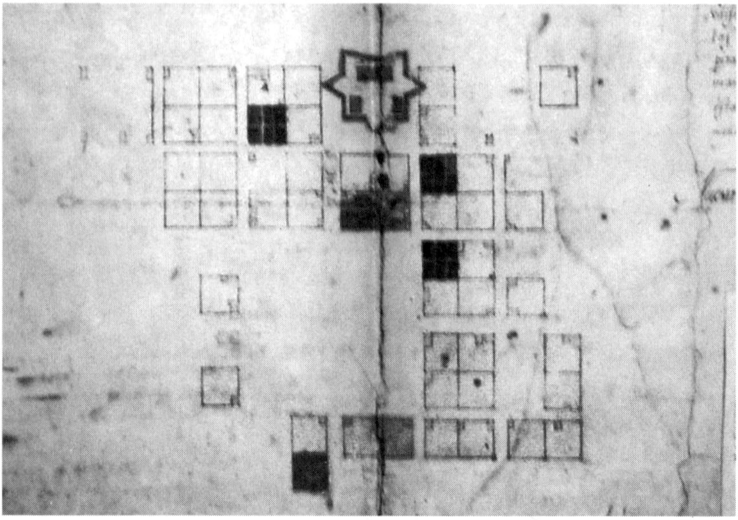

Detail of Juan Perchet's map of Galveztown from 1779 showing the fort and the layout of the town. The star shape represents the fort. *Special Collections, Tulane University*

wooden, galleried house exactly 32 feet by 16 feet; farmlands were cleared around the village.

Much too soon, however, Bouligny's concerns began to prove valid. High water on the Amite during the spring threatened construction, which had already fallen behind schedule because of a shortage of materials. The fields of corn, rice, beans, potatoes, peas, and flax were inundated. Perhaps, Collell pointedly wrote to Galvez, the governor had only visited during the dry season.

When Father Francisco Lopez arrived in early April to serve as community priest, he gathered the congregants in a makeshift chapel arranged in the corner of the barracks and led them in prayers for the waters to recede. Soon after, however, his duties began to include more cheerful occasions, including officiating at marriages between Isleño girls and Spanish soldiers.

Day-to-day life was arduous, and the Spanish colonial government was too financially strapped to respond quickly to Galveztown's ongoing needs. The litany of woes resembled biblical plagues—flooding, hurricanes, drought, and disease, as well as lack of food, materials, and tools. "I am so sad I cannot do anything about it," Collell wrote Galvez when malaria and smallpox overwhelmed the village, transforming many homes into hospitals and leaving fields untended. Yet the settlers had somehow constructed the fort, cleared the plaza, built houses, and cleared additional land for planting.

The establishment of Galveztown followed soon after Spain's declaration of war against Great Britain and not long before Bernardo de Galvez marched on Forts Bute and New Richmond. And, since Galveztown was seen as a second front to protect Spanish interests and fight the British, Galvez ordered the village militias to attack British ships and posts on the Amite River. Among their targets was Fort Graham near Canewood, the settlement from which the Galveztown squatters had come.

The militias enjoyed reasonable success: they captured seven British ships and more than a hundred sailors and soldiers, among whom were "3 Serjeants. 1 drummer, 7 rank and file Independent Company."

After Galvez's victory at Mobile in 1780, West Florida had been reclaimed for Spain, and Galveztown had no further strategic importance. Yet it survived. In the summer of 1781, Captain Gilbert Antonio de St. Maxent replaced Collell. St. Maxent arrived at his new post brimming with self-confidence, ready to transform pathetic Galveztown into a successful community like Valenzuela, another Isleño settlement located on Bayou Lafourche, where he had been commandant.

But St. Maxent's luck did not cross the Mississippi River with him. Floods continued to destroy the crops, Indians raided the little village, and morale sank so low that St. Maxent was forced

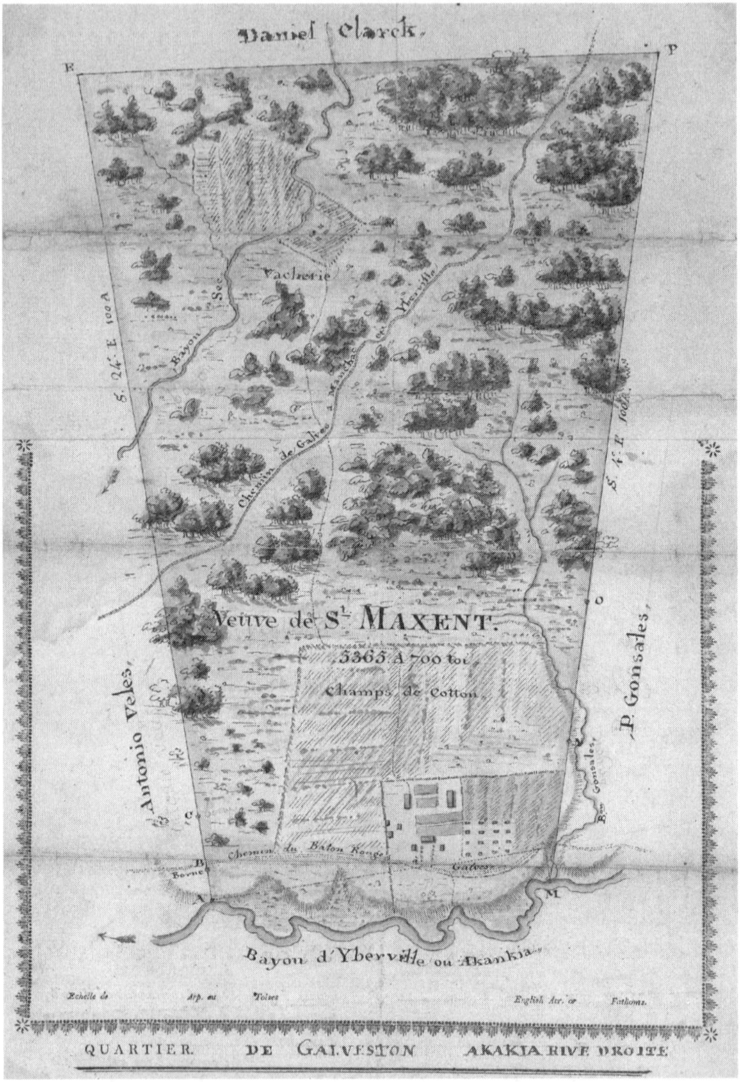

Barthelemy Lafon's map of the Widow St. Maxent's Galveztown property, 1804. *Barthelemy Lafon Land Survey, Mss. 4821, Louisiana and Lower Mississippi Valley Collections, LSU Libraries, Baton Rouge, La.*

to forbid colonists from leaving. To cajole their loyalty, Governor Estevan Miro, Galvez's successor, offered residents the opportunity to sell their surplus produce in New Orleans. This offer presumed that the 242 residents who remained at Galveztown could somehow raise enough crops to spare any for sale.

By the time the defeated St. Maxent departed in 1787, the village had also lost its priest. Church services and sacraments were performed only when the priest from San Gabriel could travel to the settlement. In marked contrast to the experience of San Gabriel, the loss of the Galveztown church was one more hardship that provoked settlers to abandon their houses and fields and seek better lives elsewhere.

Three hurricanes in 1793 and 1794 and flooding from the Amite River further undermined the poor living conditions. By 1796, only twenty-one families, barely surviving, clung to their Galveztown properties; by 1805 nearly all had departed. Some moved to Baton Rouge near the fort, in the area now known as Spanish Town, in the shadow of the present state capitol. Others drifted south and inland to higher ground, eventually establishing a new settlement named Galvez.

THE VOYAGE OF THE *NONPAREIL*

The tiny remnant population left at Galveztown was no doubt surprised when, in June 1805, a sailing ship appeared that had, remarkably, fought its way down Bayou Manchac from the Mississippi. Somehow, the bayou had retained its unfounded reputation as a thoroughly navigable waterway, a shortcut from the Mississippi to New Orleans, and Captain Jonathan Devoll of Marietta, Ohio, had been lured by this fallacy.

Devoll had built the small, light-draft schooner, the *Nonpareil*, himself. The magazine *American Pioneer* noted that it had "a large cabin to accommodate passengers, was beautifully painted, and sat on the water like a duck in her native element." The captain had confidently set sail down the wide open Ohio River into the Mississippi and turned innocently into Bayou Manchac to follow what he believed to be the speediest route to New Orleans, where he planned to set up a shipping and transportation service on Lake Pontchartrain. As the magazine reported, the men "were assured by those who pretended to know that there was but one obstruction [along Bayou Manchac] about a mile from the entrance. When this was passed, they would find no other for the eighteen miles to Galveztown."

The *Nonpariel* entered Manchac on June 3 with her large square sail hoisted. "Her ill-fated crew commenced their Herculean labors of opening a passage through multiple obstructions which had closed the way," the *American Pioneer* reported. "Day after day was consumed with constant labor with axes and saws." They made headway at all only because a passenger who had boarded at Natchez had fortuitously brought along crosscut saws. Without him, "it is very doubtful they would have extricated the vessel from this difficult pass." Nor could they have turned around; the force of the Mississippi River current in the bayou was too strong to sail against.

The *Nonpareil* fought its way for twenty long days, much slower progress than even the frustrated Iberville had made. Among the crew's diversions was observing the impressive population of alligators. "This bayou being little frequented by man," reported the *American Pioneer,* "the alligators live here in peace and quietness," save for Indian and Spanish hunters.

When the *Nonpareil* finally reached Galveztown, the crew sold what was left of their provisions to the remaining residents—"meal for $3 per barrel and the remainder of cheese at 50 cents a pound, and a few barrels of flour at $14." Captain Devoll took on a pilot to help navigate the rest of the trip, not realizing the worst was behind him.

American geographer and surveyor William Darby arrived at the Galveztown location in 1816 to find what was left was "of little consequence . . . a few ruined houses." Seemingly ignorant of its history, however, he added optimistically, "This town must in the progressive advance of Louisiana become a place of note. . . . Its situation gives it many natural advantages that will in no great length of time be called into operation."

Galveztown remained a dot on Louisiana maps for a long time after it ceased to exist as a community. The Graham/Tanner map of 1834 shows Galveztown as if it were still a settlement of significance. Perhaps the village was noted as a landmark because of rumors that legendary pirate and smuggler Jean Lafitte had hidden part of his treasure there. (Nothing was ever found to support the story, however, which likely arose because, after participating in the Battle of New Orleans in 1814, Lafitte moved to Galveston, Texas, where he lived out the rest of his life.)

In a promotional pamphlet to sell Ascension Parish land in 1888, Galveztown was mentioned: "Not a vestige of the ancient town remains save here and there small bits of brick and mortar which the trickling rain or burrowing snouts of swine unearth to bear mute evidence that here was once a brick-paved city, said to have contained a thousand inhabitants or more. The spot is today the sole property of a lineal descendant of the brave explorers who founded the town. Mr. Miguel Gonzales, who has attained more than fourscore years, is the lone inhabitant and owner of the site of this ancient city."

Today the site of the old village is buried under the silt of multiple floods, road construction, and contemporary development. A historical marker and the pride of Glenn Cambre and a few others like him are all that remains of a village that fleetingly presided over the junction of Bayou Manchac and the Amite River.

10 What the Louisiana Purchase Meant

AFTER BERNARDO DE GALVEZ'S CONQUEST of British West Florida in 1799, Bayou Manchac remained a feature on contemporary maps, although it was no longer an international boundary and the back route had lost its allure. In 1802, however, political turmoil in Europe and uneasiness about the ultimate intentions of the United States combined to provoke Spain to cut off the fledgling nation's access to New Orleans and the lower Mississippi. Angry residents from the American west—Kentucky, Ohio, western Pennsylvania, and Tennessee—threatened to resort to force to reclaim access to the vital port city.

In the meantime, after a difficult war with France, the Spanish government had negotiated a secret retrocession of territory, returning the Isle of Orleans and Louisiana to their original claimant. Emperor Napoleon Bonaparte had long wanted to restore his country's position in North America; he should have been elated. But it soon was evident that the immense colony would impose an immense financial and political burden that France could not support; its resources were stretched thin in response to provocations from Austria and England and the demands of dealing with the slave rebellion in their Caribbean colony of Saint Domingue.

In 1803, American president Thomas Jefferson dispatched Robert Livingston and James Monroe to Paris to try to purchase the Isle of Orleans and so gain unimpeded use of the port of New Orleans. Livingston had previously, and unsuccessfully, tried to buy New Orleans after the United States had learned of the secret retrocession. But now, Napoleon's treasury minister, the Marquis de Barbe-Marbois, made an astonishing offer: France would sell not only the Isle of Orleans but also the vast, unmapped expanse of Louisiana—the unknown lands that France had formally claimed after Iberville's journey. The price was, ultimately, $15,000,000 for what was found to be almost 828,000 square miles. Jefferson chose William C. C. Claiborne, former governor of the Mississippi territory, to oversee transfer of Louisiana to the United States and then to serve as governor of the Territory of Orleans.

This extraordinary transaction infused Napoleon's coffers with needed monies and proved a remarkable boon to the United States. The land acquisition strengthened the status of the American republic in relation to Great Britain, which controlled Canada, and offered the opportunity of great westward expansion as well as securing the Mississippi River.

Among its most unnoticed effects was that Bayou Manchac had once again become an international boundary. Now, it formed the border between the United States and Spanish West Florida, which was not included in the Louisiana Purchase.

The restoration of the bayou as an international boundary may have influenced its selection as the site of a famous duel between two New Orleans political luminaries in 1807. Governor Claiborne and Mr. Daniel Clark, delegate to Congress from the Orleans Territory, met on the West Florida bank of the bayou to settle an enduring hostility.

Clark was an Irish immigrant with a reputation as a dominant, complex, difficult man. He had been appointed regular consul in New Orleans in 1801 and was disappointed and resentful when Claiborne, not he, was named governor. The *Louisiana Gazette* reported that Clark had said that "the militia of the Territory had been neglected and had seen a black corps preferred to them." This constituted an insult to Claiborne, who demanded a retraction from Clark. Clark refused.

The men engaged in insults and affronts for years, and both gossip and history have suggested assorted reasons for their duel. But whatever the ultimate provocation, it compelled the enraged Claiborne to challenge Clark to settle the matter.

The Old World tradition of dueling, a gentlemanly engagement, was still lawful in New Orleans, but perhaps the two men chose to meet across the bayou because it was neutral territory, beyond Claiborne's jurisdiction, and less conspicuous than the environs of New Orleans.

The men and their seconds arrived with pistols on Monday, June 8, 1807, near where old Fort Bute had stood. Their weapons—pistols—represented a very American choice; Creoles would have preferred dueling swords—*colichemards*. The south Louisiana day would have been warm and humid as the four men gathered in a muddy clearing surrounded by a stand of great forest. The duelists were probably clad as gentlemen of the day would have been—in knee breeches, white linen shirts, and boots, having shed their traditional vests, coats, and hats to be battle ready.

Shots rang out almost simultaneously from both pistols.

In a letter to a friend after the duel, Clark offered his version of the action: "We fired almost at the same instant, at ten paces and the Governor fell, shot through the thigh and with a most severe contusion on the other. I have received no injury . . . and will return to Orleans in three or four days," perhaps after a brief respite at his Mississippi River plantation.

Governor Claiborne, in a letter dated June 16, acknowledged, "I have suffered much pain, but the wound has assumed a favorable aspect. I hope in ten to fifteen days to be enabled to walk." The New Orleans French-language newspaper, *Le Moniteur,* noted that the governor was doing "as well as could be expected" but would not be well enough to appear in public for a month after the incident.

After the duel, the two men remained adversaries. In 1809, after having chosen not to run for a second term as congressional delegate, Daniel Clark backed an ally for the position who lost to Claiborne's friend Julian Poydras. When Clark died in 1813, he was buried in St. Louis Cemetery; it was, ironically, where two wives of Governor Claiborne were also interred.

Upon completion of the Louisiana Purchase, Thomas Jefferson argued that West Florida had logically been part of the acquisition. As part of Spain's retrocession to France, it had therefore been included in France's agreement with the United States, Jefferson maintained. But the Spanish crown disagreed: West Florida was a separate territory. Thus the lands on the upper bank of Bayou Manchac remained in Spanish hands after 1803.

Spanish rule was benign. Residents of West Florida not Spanish by birth—primarily British and Anglo-Americans—had acknowledged a convenient allegiance to Spain in order to retain their lands. English and French were still allowed as official languages. After the Louisiana Purchase, however, West Florida was surrounded by the United States; the fort at Baton Rouge was the sole remaining Spanish post on the Mississippi River. And, to no one's surprise, in 1809 the United States placed trade restrictions that curtailed the use of the lower Mississippi by the colony of West Florida.

A committee of West Florida leaders with distinctly Anglo names—Samuel Fulton, Philip Hicky, Joseph Sharp, and John Davenport—met with Spanish authorities to promote opening Bayou Manchac as an alternative trade route. As the British had done forty years before, the committee recommended removing driftwood, trees, and other obstacles from the first segment of the channel and creating a ninety-foot clearing along its bank, a roadway that could double as a towpath during low water.

These improvements plus a few bridges would be of minimal cost, they argued. And for a slightly larger investment, engineers could also dig a feeder canal from the Mississippi into the bayou, to "give such force to the waters of Manchac that in a few years that channel would be navigable for large ships." It was remarkably similar to Lieutenant Philip Pittman's proposal of 1765 that the British had never implemented. The West Florida authorities deemed it possible.

But the work never began. A growing unrest had been brewing among West Florida's citizens, especially the pro-American faction, who were sorely disappointed when they had been excluded from the Louisiana Purchase. They plotted secretly to force the transfer of West Florida to the United States and sprang into action on September 23, 1810. A group of seventy men attacked the fort at Baton Rouge and, in a brief battle, easily captured the fort. The leaders declared the former colony of Spanish West Florida to be the new and independent West Florida Republic. They wrote a constitution and established a new government for their new nation, even as they applied to President James Madison for annexation to the United States.

Madison's response was swift. He was as anxious as Jefferson had been to claim West Florida and dispatched Governor Claiborne, long recovered from his dueling wound, to Baton Rouge.

On December 10, 1810, after only seventy-four days of independence, the West Florida Republic became the county of Feliciana in the United States of America.

Some historians have speculated that the rebellion was plotted and carried out by mercenaries planted and paid for by the United States government. Certainly there had been no secret about American interest in acquiring West Florida. But the composition of West Florida's residents, which included successful landholding former British and Anglo-American citizens, seems to indicate that the rebellion was primarily homegrown. And, in their declaration of independence establishing the new republic, its leadership specifically expressed the wish to become "an integral and inalienable portion of the United States."

Despite its short life, the West Florida Republic became part of the history and lore of the state of Louisiana, and the flag of its ephemeral government—pale blue with a white star at its center—flies overhead wherever the ten flags of Louisiana wave.

But the demise of the West Florida Republic once again demoted Bayou Manchac from its status as an international boundary. The winding eighteen-and-a-half-mile stream—partly dry, partly navigable—became instead just another small, American waterway. After Louisiana's admission to the Union in April 1812, Manchac was merely a stream separating several parishes of the new state.

11 When Andrew Jackson Dammed the Bayou

THE WAR OF 1812 BROKE OUT a month after Louisiana became a state. The United States had declared war against Britain in response to numerous provocations, including the blockading of ports, hijacking of American merchants on the high seas, and inciting Indians to attack western settlements. Soon after the war began, the British laid siege to their former posts at Pensacola and Mobile before turning, inevitably, to attack New Orleans.

The port city was key to American trade along the entire Mississippi River valley. Should it fall to the British, they would reclaim vast lands in North America, and the United States would suffer. In 1814 the United States Army dispatched General Andrew Jackson to New Orleans to defend the city.

Jackson was a politician and a soldier, recognized for his uncompromising nature and enormous self-confidence. Fresh from victories in other battles, he arrived in New Orleans to confront the approach of the British Royal Navy, although no one knew which route they would take. They might sail upriver from the mouth of the Mississippi or into Lake Pontchartrain and through one of several bayous that connected to the river. Or they might descend from a point north of New Orleans along the Mississippi.

Jackson, unfamiliar with the region, ordered the damming of all south Louisiana bayous that flowed into or out of the Mississippi, to avert any possible secret approach. One of these, of course, was Bayou Manchac.

It is unknown whose advice Jackson was following. Some believe it was given by General David Morgan, with whom Jackson stayed overnight on the north shore of Lake Pontchartrain on his way to assume his New Orleans command. Others suggest the source was the pirate Jean Lafitte, who had been pursued by the territorial government under Claiborne, and had been approached by the British to participate in the attack on New Orleans. (He declined and declared his allegiance to the United States.) But it is known that Lafitte met with Jackson on December 4, 1814, to offer his services and those of his men in defense of the city.

Most likely, however, the suggestion to barricade the bayous came from Arsene Latour, an engineer recommended to the general by Robert Livingston. Latour was reputed to be "a man of perfect honor and integrity who speaks both French and English fluently" and who could be helpful to Jackson in rallying the Creoles of New Orleans to defend the city. Furthermore, Latour had assisted in mapping the area and informed Jackson of seven different routes the British might take (none of which was by way of Manchac).

Jackson ordered a dirt levee reinforced with whole cut trees to be erected across Manchac at the Mississippi. It was an emergency measure and intended to be temporary.

Given Manchac's natural hydrology, of course, damming it was an excessive precaution. After the difficulty that the *Nonpareil* had encountered nine years earlier during high water and knowing that the channel from the Mississippi to the Convergence had never been cleared, local residents would have laughed. Manchac was too great an obstacle course for warships. Furthermore, it was unlikely that the bayou would have been sufficiently flooded in November and December when battle preparations took place, since flooding usually occurred between March and June. Given normal flood cycles, Manchac's bed would have been considerably higher than the Mississippi at that time of year, impossible to reach from the river without a lock.

The Battle of New Orleans took place south of New Orleans on January 8, 1815, far from Bayou Manchac, but Jackson's defensive strategy had a far-reaching and long-lasting effect. His temporary dam protected thousands of acres near the bayou, rendering them less susceptible to annual flooding and more productive for planters. It also diminished the threat of flooding along the backlands of the bayou. Area landowners loved the dam.

IS A BACK ROUTE POSSIBLE?

Damming the bayou was not a new idea. It had first been officially suggested—for political, not military reasons—in 1807 when the Legislative Council of the Territory of Orleans recognized the severity of flooding and asked landowners to petition the president. Nothing came of the petition, but in 1826, after living a dozen years with Jackson's dam, a consortium of landowners approached the state to make the closure permanent. The governor objected and asked the state legislature to petition Congress to remove the dam, to reopen the bayou.

But local politics won out. The Louisiana legislature passed an act mandating permanent closure of Bayou Manchac at the Mississippi and appointed five advocates of the dam to a supervisory commission to oversee construction of a well-engineered, durable levee.

Settlement had already increased along the fertile high ground bordering the Mississippi between New Orleans and Baton Rouge. Plantation properties—expansive tracts growing sugarcane below

Manchac and cotton farms above the bayou—had begun to line the river, although the planters faced many of the same challenges that had overwhelmed the residents of Galveztown. Spring floods along the Mississippi could often break through their levees, causing a crevasse,* which could inundate the countryside far inland, occasionally as far as the Amite River.

The natural levee created along the Mississippi by annual overflow had been augmented by a system of man-made levees, in place since French settlement. Each landowner was responsible for constructing and maintaining the levee along his own riverfront. But entrances to the distributaries had never before been closed. Damming Manchac was unprecedented.

Of course, the landowners' satisfaction with the dam erected at Manchac did not consider how it might ultimately affect the natural hydrology of the Mississippi River system, which depended on its distributaries to serve as outflow channels. Opinion favored the concerns of the settlers along these waterways, who demanded as much protection as possible.

In the nineteenth century, levees were the preferred method of Mississippi River flood control, based on the premise that levees would protect the lands behind them and also increase the force of the current flowing between them, to scour and deepen the channel. This would mitigate the effects of annual flooding. Or so the prevailing argument held until the great flood of 1927 disproved it.

Without the natural flow of water and dispersal of sediment, however, swamps and the alluvial plain suffered. In the Bayou Manchac area, higher Mississippi River levees plus a blocked Bayou Manchac began to cut off replenishment in the natural "catch basins" of Spanish Lake, Bluff Swamp, and nearby lowlands.

* A crevasse is a break in a levee caused by some underlying weakness through which torrents of water rush with such great force as to gouge nearby land and flood vast acreage, often destroying buildings and deluging crops, machinery, and animals.

If some landowners delighted in the dammed bayou, other residents of the area never abandoned the dream that had begun to percolate with Iberville, namely transforming Manchac into a completely navigable stream, to create a useful back route. A succession of proposals, schemes, and studies dedicated to this end persisted for over a century.

The first of these arose in 1835 when champions of the back route approached the state legislature and succeeded in having a resolution accepted to investigate removal of the now-permanent dam. The investigation led nowhere. Proponents then turned to the United States Senate, which responded by commissioning a study. The result came in an official report to the Secretary of War in 1842, signed by T. B. Linnard, a member of the Corps of Topographical Engineers:

> The head of the bayou consists of a dam of earth in continuation of the levee, wide enough for a carriage way for which it is being used. The bed of the bayou between the levee and the river, about 100 yards, has been filled up to within five or six feet of the banks by deposits made during high water. . . . The cost of removing the obstructions would not exceed fifty dollars; a narrow trench cut through the deposit outside of the dam and an opening at high water in the dam itself would be sufficient. . . .

But their evaluation was pragmatic:

> The dam is not the only obstruction to navigation. Manchac for six miles from the Mississippi is extremely crooked, has an average breadth between the tops of the banks of about 14 yards and is nearly dry. The banks and the bed are grown up with young timber. [Beyond the Convergence] Bayou Manchac becomes thirty yards wide and from 7–10' deep. From Fontaine [*sic*] to the confluence with the Amite, there is much fallen timber in the stream and many leaning trees on the banks. This is the only navigable part

> of the bayou. It may be cleaned out and made practicable for schooners at a cost of $6000, which includes removal of timber from banks for a space of forty feet breadth.
>
> If the Mississippi water were admitted freely into Manchac, the lands on [Alligator Bayou and Bayou Fountain] would greatly depreciate in value and the most serious injury to the plantations on the river south of Manchac would ensue. The navigation would be of no public utility if all the trees, logs, and stumps were cleared out [because] no boats of any description now used on the Mississippi could pass through the upper portion. . . . The bed of the bayou at its head is about 15' above low water mark in the river, so no water could enter for several months. . . . I would respectfully recommend that the obstruction at the former entrance from the Mississippi into Manchac be not removed.

This succinct and most practical report, however, failed to kill the dream. On several occasions during the 1850s, *DeBow's Review*, a popular southern business journal, propounded the advantage of reopening the bayou. "This having once been a natural issue or branch of the Mississippi would form a far better safety valve for the escape of surplus waters in time of a flood than all artificial ones," editorialized J. D. B. DeBow.

Scientifically, he was correct. Politically, his proposal would have been stronger before the land had been settled. Now, however, there were greater financial implications for those with property if the dam were removed. A letter to the editor of *DeBow's* by Philip Winfree rebutted the editorial. Winfree, however, somehow neglected to identify himself as a member of the commission that had overseen the permanent dam's construction in 1826. "Now this bayou never was fairly open to the Mississippi," he wrote, not inaccurately. "And if it had been, it would have inundated the whole island, of course, including the city of New Orleans. . . ." This was certainly hyperbole.

In 1862, during the Civil War, the *New York Herald* editorialized in favor of opening up Manchac on behalf of long-neglected national defenses. "In time of war," the paper intoned, "an interior coast channel from New Orleans to Boston can be completed at very small expense." Part of this national artery would include "opening up the bed of the Iberville River and inner communication may be continued along the coast . . . via Mobile and Pensacola" and up the Atlantic seaboard. This grand dream also included a trans-Florida ship canal.

During the same year, "the great crevasse of 1862," as *DeBow's* called it, occurred along the Mississippi not far above Bayou Manchac. Water poured into the dry segment of the bayou, inundating all lowlands from Manchac north toward Baton Rouge and inland for two and a half miles. Such an event only heightened concern about what could happen if the bayou were permanently reopened.

During Reconstruction, the federal government considered opening Bayou Manchac. Their motive: to introduce first-class steamboat navigation along the back route and help jump-start regional economic development as Louisiana and the South attempted to recover from the devastation of the Civil War. First Lieutenant J. K. Hezlep of the Army Corps of Engineers conducted a survey to determine the cost and efficacy of building floodgates and locks at the Mississippi and clearing the channel.

Hezlep reported that the land along both Bayou Manchac and the Amite River was bordered with heavy forests and few plantations or landings. "In many places the trees overhang the bank, so as to render navigation difficult and dangerous," he noted. The first segment of the bayou was filled with logs, trees, and undergrowth except when excessive flooding from the Amite had forced water into the first mile or two west of the Convergence. Many places below (east of) Alligator Bayou could be under more than twenty

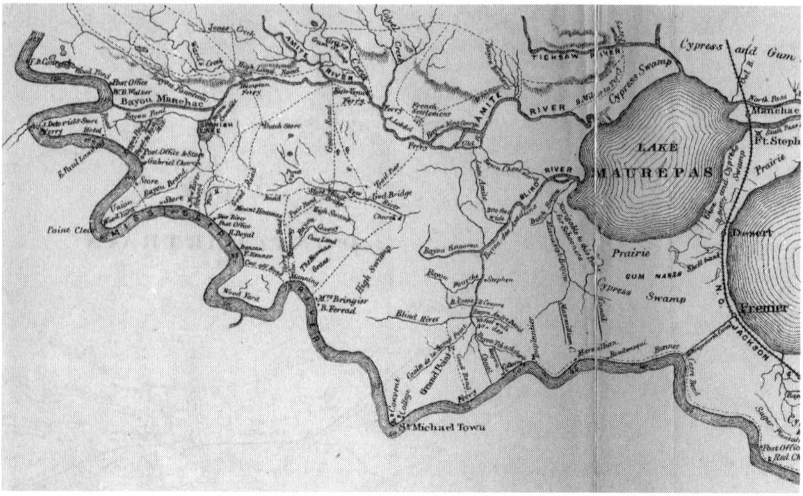

Detail of map prepared in 1868 by Lieutenant J. K. Hezlep, showing Bayou Manchac area as of that year. The bayou is in the upper left quadrant. *Collection of the Louisiana State Museum*

feet of water and, "overflowing a large portion of the country on the south side and falling rapidly, a very rapid current is formed rendering navigation extremely difficult."

To overcome Manchac's peculiar hydrology, Hezlep recommended a system of four locks—a main lock and three side locks—to be constructed at the Mississippi, opening sequentially depending on the river stage. He also recommended widening the bayou to forty-five feet "to permit boats (which must be sternwheelers) to pass each other readily at low speed."

Hezlep's plan reflected a more sophisticated understanding about the river and knowledge of engineering technology than had Pittman's proposals of the previous century. Hezlep understood that the Mississippi current would shift the riverbed at Manchac and undermine a lock; he realized that crevasses could occur anywhere, that the Manchac channel must be deepened, and that steamboats able to cruise the Mississippi and Bayou Manchac were impractical for sailing in the open water of the lakes and the Gulf of Mexico.

At the end of this litany of challenges, Lieutenant Hezlep could not recommend the project. "The channel proposed will be less useful and convenient than many others that might be selected [from New Orleans through the lakes]. . . . [I]t will present about 75 miles of intricate navigation between the river and Lake Pontchartrain, requiring a speed of 5 mph, taking about sixteen hours."

But *DeBow's Review* was relentless, printing a transcript of a speech to the St. Louis Board of Trade. The presentation wishfully envisioned that "ordinary western river steamers may load at our wharf [in St. Louis] with produce, especially grain in bulk, manufactured goods of all descriptions turned out of St. Louis workshops and foundries, and land alongside vessels anchored in the deep waters of Mobile Bay and discharge into them the shipments made from here to the Atlantic seaboard or the market of the old world . . . at a rate so cheap our merchants could undersell all western competitors." It was surely not coincidental that St. Louis was then engaged in a fierce trade competition with New Orleans.

In 1874, Erastus Everett, writing in *Appleton's Journal,* advocated reopening the bayou, but for flood control rather than commerce. "The bayous were closed by a policy as selfish as it was shortsighted," he wrote. Bayous such as Manchac, he opined presciently, were "safety-valves, through which the floods were poured . . . conduits to convey surplus waters to the ocean during the spring floods when they rushed with appalling impetuosity. . . ." Therefore, he argued, the large bayous should be "reopened, made of suitable depth and protected by levees."

Again, the argument was scientifically correct, but no one listened. The next action by the Congress of 1879–80 appropriated funds only to clear the navigable portion of Bayou Manchac.

THE ATTEMPTED VOYAGE OF NATHANIEL BISHOP

The various arguments about Bayou Manchac must never have come to national attention, however, because the myth of its navigability persisted. And Nathaniel Bishop, like Captain Devoll before him, believed it.

In 1875, Bishop, a grower of cranberries and sailing enthusiast from Toms River, New Jersey, decided to abandon his daily routine and travel down the Ohio and Mississippi. He built the *Centennial Republic,* a Barnegat sneakbox cruiser, a type of sailing craft popular in the mid-Atlantic states.*

Bishop's expressed plan was to "follow D'Ibberville's ancient route to the sea. . . . The statement is repeatedly made [in contemporary guidebooks] that steamboats can go through this bayou into the Amite River and down," he wrote in his memoir, *Four Months in a Sneak-Box.* So when he had sailed as far as Baton Rouge, Bishop pulled ashore to seek information about how to gain access to the bayou that the French explorer had traveled.

When he arrived at Baton Rouge, he had "asked a local crowd gathered at the levee the chances of a passage through Bayou Manchac. . . ." One man advised him to seek help at Walker's plantation,

*"The sneak-box offered ample stowage capacity . . . to carry the amount of baggage for the voyage; for I was to avoid hotels and towns, to live in my boat day and night, to carry ample stock of provisions, and to travel in as comfortable a manner as possible. In fact, I adopted a very home-like boat, which, though only twelve feet long, four feet wide, and thirteen inches deep, was strong, stiff, dry, and safe; a craft that could be sailed or rowed as wind, weather, or inclination might dictate,—the weight of which hardly exceeded two hundred pounds,—and could be conveniently transported from one stream to another in an ordinary wagon." From chapter 1, *Four Months in a Sneak-Box,* by Nathaniel Bishop.

where field hands would help lift his boat over the levee "'and put it into Bayou Manchac which is about 100 yards from the banks of the Mississippi. Its mouth was filled up a long time ago but when once in the bayou you can float to the Amite and so on to the Gulf.' Another voice contradicted: 'why, the bayou is dried up for at least eight miles at its head.'

"Another man intruded. 'I live on the Bayou Manchac and can assure you that after you have hauled your boat through Woodstock Plantation of the Walker family, you will find enough water in the bayou to float down upon to the Amite River.' The crowd began to discuss and contradict each other," and Bishop departed to seek the truth for himself.

"I soon found the accomplishement [*sic*] of my plan impossible as the dry bottome [*sic*] of the bayou was 15' above the water of the Mississippi," Bishop wrote with exasperation.

When he drew ashore at Woodstock to ask permission to find and follow the bayou, Bishop learned from the plantation owner that the bayou was in fact dry for a considerable distance: "We walked to a shallow depression near the house; it was dry and carpeted with short grass. 'This,' said Mr. Walker, 'is the Bayou Manchac which D'Ibberville descended.'"

One of Walker's field hands offered himself and three men to help haul the sneakbox over the levee and then cart it by wagon to open water where "sloop navigation commenced, at a point called Hampton's Landing." That place was eight or nine miles inland.

But his romantic notions of following Iberville's passage were dashed, and Bishop declined the offer. He had wanted to sail the bayou, not jounce alongside it in a wagon. He continued his voyage along the Mississippi and later published the engaging memoir about his journey. But posterity was deprived of a vivid first-person account of a mid-nineteenth-century voyage on Bayou Manchac because, of course, the back route did not exist.

THE DREAM LIVES ON

In 1888, Congress appropriated funding for removal of "snags, logs, and impending trees between the mouth [of Manchac] and Hope Villa," a small community that had grown up on the bayou at Hampton's Landing. But the order again omitted any reference to opening the full length of the channel.

Clearing the bayou was apparently not a routine chore, however, because a 1910 headline in the daily *New Advocate* of Baton Rouge cheered the Corps' reappearance on the bayou. "Cleaning Up at Manchac," screamed a large headline over an article that welcomed the "stump snatcher and log remover boat Pearl." The article noted that "the bayou has been well nigh closed to navigation by the accumulations of logs, sinkers, stumps and trees," which impaired boat traffic along the navigable portion—steamboats that sailed from New Orleans through the lakes and barges hauling local cypress. In the same newspaper, a complementary editorial gibed, "One of the foolish things [Andrew] Jackson did was close the head of the bayou"—which seemed to indicate that the dream of the back route lived on.

Seven years later, when what was touted as "the model road from New Orleans to Baton Rouge" was plotted through Hope Villa, a bridge was constructed to replace the ferry that had historically carried passengers, cargo, and animals across the channel. Engineers had planned a fixed bridge but local citizens protested: the bridge must accommodate the small packets from New Orleans that serviced Hope Villa and other "boats of reasonable size." The fixed bridge became a turn bridge.

In 1928, Louisiana's Sixth District congressman, Bolivar Kemp, attempted to rally interest in clearing the first segment of the bayou. He succeeded only in provoking the Corps of Engineers to clear obstructions from the navigable segments, once again for the purpose of flood control. This may have prompted local resident and civic leader Louis U. Babin to echo Kemp's appeal but with a clever variation: Bayou Manchac should be developed as a link in the Gulf Intracoastal Waterway system.

The Gulf Coast waterway, a system of natural and man-made waterways, dates from 1905. Today it is a marine highway stretching from Texas to Florida, intersecting with navigable rivers, including the Mississippi, to form the primary commercial water route across the Gulf states.

"The bayou is but temporarily prevented from serving its rightful purpose of connecting the waters of the Mississippi, lakes, and gulf and the sea," Babin dramatically contended to the Corps of Engineers. Why not allow commercial and industrial boats from the Mississippi to follow a navigable Manchac into the Amite and the lakes to the Intracoastal Waterway?

The most positive result of Babin's proposal was the replacement of the turn bridge over Bayou Manchac with a new, fixed pony-truss bridge in 1931. It was "constructed so it will not interfere with any water traffic on the bayou," reported the *Baton Rouge Morning Advocate*. The cement and steel structure had a span of eighty-two feet, more than ample to accommodate the limited boat traffic that passed beneath it.

Bolivar Kemp reappeared in 1932 to reintroduce his proposal, but another obstacle to creating the back route had already arisen: the uniform levee along the Mississippi River was under construction in response to the devastation of the 1927 flood. With completion of this massive barrier, the only way to link the river with its distributary would be by constructing a lock, which would increase the cost exponentially and require huge political clout. Not surprisingly, a lock for Bayou Manchac at the Mississippi was not a national priority, and Kemp's measure went nowhere.

And so, as the years went by, the bayou remained dry from the

Mississippi to the Convergence and navigable only by small boats from the Convergence to the Amite. Over time, the head of navigation (or at least navigability) shifted east, from the Convergence to Ward's Creek, as the channel silted in. Periodically, the Corps of Engineers cleared and desnagged the bayou and removed fallen trees to open the channel for drainage and flood control.

In 1962, the ghost of Louis Babin unexpectedly appeared in the guise of a resolution sent to Congress by the Louisiana Senate asking that Bayou Manchac be opened for commerce. As the Baton Rouge *State Times* editorialized, "a new alternate barge canal route opened from Morgan City to Baton Rouge last summer. Tows through Manchac would make sense." Navigating through the back route would shorten the distance from Baton Rouge to the Intracoastal Waterway by 125 miles. But the editorial also wisely noted that "the cost of building a lock and extensively dredging and widening the channel from the Mississippi to the Convergence would be too expensive to justify." The estimated price tag had grown to $4 million.

Another, quite novel proposal emerged in 1970: a plan to convert Manchac Bend, the broad curve of the Mississippi from which the bayou originally flowed, into an oxbow. By cutting a diversion for the Mississippi River through Manchac Point, the Mississippi would flow straight south; its former bend would be dammed at the upper end, creating a recreational lake and a protected harbor accessible from the open, lower end. This engineered waterway might become a first step toward creating the back route.

It was a visionary but impractical scheme, which conveniently overlooked both the expense of the project and the potential dangers of inviting a dedicated recreation area and a dedicated commercial shipping channel to share such a small space.

The final dream of opening Manchac as a commercial waterway came in 1977 when the Corps of Engineers again dedicated

$10,000 to study the channel's potential for barge traffic and adding new industry along its banks. The study revealed the same problems that had always existed; now, however, it also rallied the opposition of residents who lived along the bayou—many more than in previous decades—as well as a small group of activists from the burgeoning environmental movement. Both groups pointed to Bayou Manchac's value as a natural attraction and fervently argued that the money would be better spent on pollution and flood control and to create additional recreational facilities along the stream.

They need not have become exercised, however, because the cost of the proposed project had become truly exorbitant. Clearing and dredging the channel plus constructing a lock at the Mississippi River would be expensive enough, but in addition, all infrastructure crossing the bayou would need to be reconfigured. That meant constructing new, raised bridges for Interstate 10 and Airline Highway and the four other roads, rebuilding two railway bridges, and enhancing rights-of-way for power lines, pipelines, and other utilities that crossed the stream.

After two centuries of dreams and schemes to make Bayou Manchac navigable from the Mississippi to the Amite, the dream of a commercial back route was finally dead.

An optimist might note here that, because the channel was never altered for commercial traffic, it escaped being dramatically dredged, lined with concrete, widened, or straightened. It had survived and might be forever protected and revered for its history, natural history, and scenic beauty. In an ironic twist of fate, however, the lack of man-made changes to Bayou Manchac may instead have allowed the little waterway to fade into near oblivion.

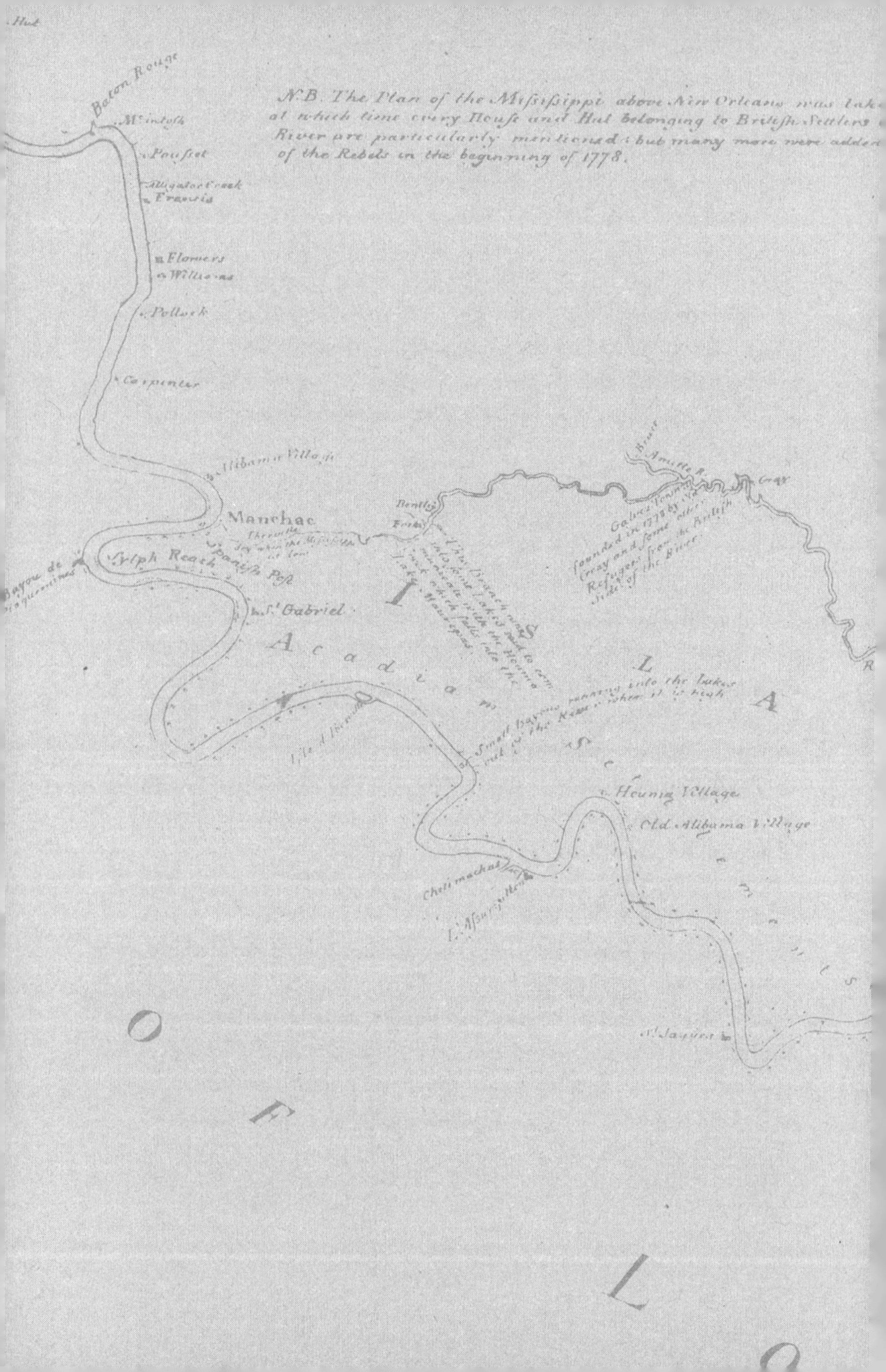
N.B. The Plan of the Mississippi above New Orleans was take
at which time every House and Hut belonging to British Settlers
River are particularly mentioned: but many more were added
of the Rebels in the beginning of 1778.
Baton Rouge
Mc.intosh
Pousset
Alligator Creek
Francis
Flowers
Williams
Pollock
Carpenter
Alibama Village
Manchac
Sylph Reach
Spanish Post
St. Gabriel
Acadia
Houma Village
Old Alibama Village
Chetimachas
St. Jaques
O
F
L

12 The Bayou in the Nineteenth Century

ONCE ANDREW JACKSON'S DAM had been made permanent, large-scale plantation properties mushroomed along the Mississippi River. By the mid-nineteenth century, the river bank was far more developed than when James Willing had raided British properties during the Revolutionary War. Vast self-contained plantation communities of elegant big houses and numerous outbuildings—kitchens, coops, slave quarters, schoolhouses, and mills—lined the river. Their extensive lands, planted with sugarcane or cotton, stretched inland for several miles.

Henry's Coast Directory, a listing of landings and properties along the Mississippi published in 1857, reflects this in the area surrounding Bayou Manchac. Manchac Plantation flanked the bayou on the lower, Iberville Parish bank and W. B. Walker's Woodstock Plantation on the upriver, East Baton Rouge Parish bank. Henry reported these as, respectively, 113 and 114 miles above New Orleans (following the river). He also noted a Manchac post office located just below Manchac Plantation.

Artist Marie Adrien Persac immortalized Manchac Plantation, owned by James Norfleet Brown, when Persac and his wife lived near Bayou Manchac with his mother-in-law. Persac's stylized rendering reveals a white, classical revival mansion, long and elegant,

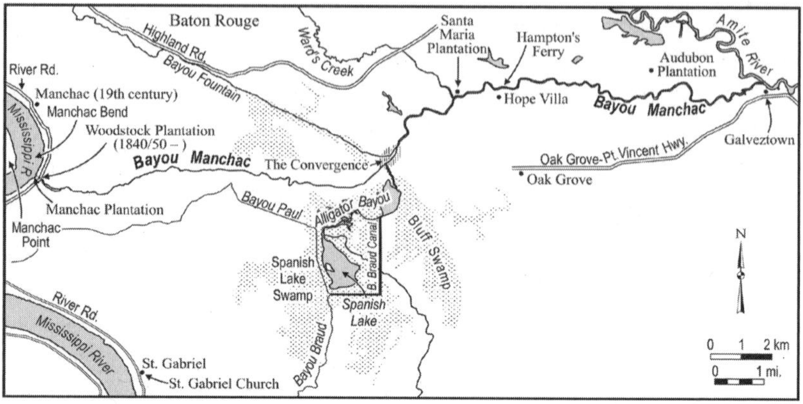

Plantations and small communities along the bayou in the nineteenth century

with a colonnaded gallery. Between the river and the house is a regimentally planted garden of mixed trees and shrubs; sheep graze in a fenced pasture while thoroughbred horses prance within another enclosure. Beyond the big house, a carriage house and stable are clearly visible, with other outbuildings including a sugar mill in the distance. The Brown family owned the property until 1899.

By the 1970s no original Manchac Plantation buildings survived, and today, a comfortable frame house and modern farm buildings are nestled within rural pastureland. In the backyard of the house, a children's swing set overlooks a quiet lagoon that defines the property line. The lagoon is a section of the first segment of Bayou Manchac, cleared of all obstacles.

Across the bayou was Woodstock Plantation, which also boasted an antebellum, classical revival big house. It was even more ornate than Manchac's and surrounded with a full complement of outbuildings. Woodstock, known as one of the most productive sugar plantations in the region before the Civil War, was owned by the Walker family from 1843 until 1899. (It was the original Mr. Walker who showed Nathaniel Bishop the "shallow depression" of the dry Manchac channel.) The Woodstock plantation house was destroyed by a Mississippi River flood sometime during the mid-1890s.*

Today, what was Woodstock Plantation is a sweep of open pasture that sprawls unimpeded from the River Road as far as the eye can see; a sturdy line of tall trees defines its Manchac border. Part of the plantation's former cane fields has been developed as a golf course and subdivision. For almost two miles, the dry channel of Bayou Manchac passes beneath rear windows and back fences of suburban homes and past the clipped green fairways of the University Club, a wild, sunken linear park filled with cypress and tupelo gum, palmetto, and shrubbery.

A bayou without water would seem an unlikely amenity to sell real estate, so I was surprised to encounter the following in an advertisement for a house in the subdivision: "[A] master bedroom porch overlooks Bayou Manchac. . . . Chef's kitchen and generous-sized keeping room . . . opens to a rear porch with views of Bayou Manchac." The eventual buyer may have been surprised to discover that this bayou attraction is always dry.

Past the rear boundary of the golf development, a narrow road parallels the winding channel along the Iberville Parish bank, steadfastly following the identifiable overgrown depression, which is crisscrossed by fallen trees and nets of dried branches. It passes through a rural landscape of continuous but well-spaced homesteads, some derelict, some old and comfortable, a few modern and elegant, almost all surrounded by broad open pastures. Some families have lived here for generations, raising cattle, catching

* The location of Woodstock was on or near land that Manchac merchant John Fitzpatrick had identified as extremely flood prone because of its position on the outside of Manchac Bend. Here, the river has constantly pushed east against the bend, the strong current continually undermining the bank. As a result, it is estimated that the river consumed approximately one half mile of land between the late eighteenth century and the construction of the uniform levee, which brought this erosion under better control.

crawfish in the bayou, playing in the channel on the rare occasions when it fills with water.

Intermittent dams of dirt fill—private driveways—lead from the road to pastures and homes across the bayou. They are constructed around sections of giant pipes, tall enough to stand in, placed to expedite passage of floodwater that occasionally pushes into the channel from the Convergence. On the rare occasions when such flooding has occurred, this part of the first segment resembles the bayou that so frustrated Iberville.

The road continues east to the Convergence where, of course, Bayou Manchac is magically full of water.

ALONG THE BAYOU

Unlike the plantation properties fronting the Mississippi, most nineteenth-century tracts along the bayou were small, although they, too, were defined by the arpent survey. But development was sparse. When William Darby traveled the bayou in 1816, he noted, "There is settlement upon the Iberville but not extensive. . . . Galveston [*sic*] is the only place . . . yet known as a village . . . consisting of a few ruined wooden houses."

Fifty years later, according to Lieutenant Hezlep, little changed. Much of the land was still uninhabited, and the bayou was bordered "by heavy forests with the exception of a few plantations and landings." One of these was advertised in a contemporary local newspaper: "Land for sale: 320 arpents, situated on a navigable waterway, richly timbered and above any overflow, is fit for the cultivation of sugar or cotton and by the non-interrupted navigation to New Orleans."

The scattered residents along the navigable segments of the bayou would have traded with and traveled to New Orleans by

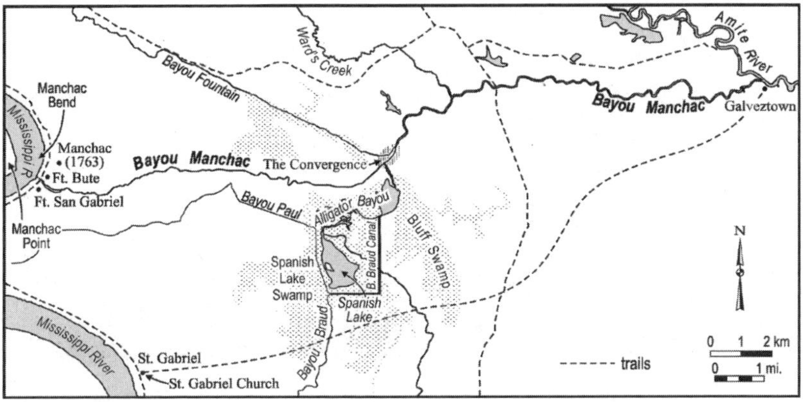

Nineteenth-century trails near the bayou

way of the Amite River and the lakes. To reach Baton Rouge, they would have followed one of the few roads—trails, really—carved through the wilderness.

One such road ran parallel to and half a mile north of Bayou Manchac. Shown on maps as early as 1817, it followed an old trail along a natural ridge from the Convergence to the Amite River. First known as Clark's Ferry Road (because it led to Clark's ferry across the Amite River) it was later renamed Clark's Road after the ferry no longer existed. Eventually it became (and has remained) Hoo Shoo Too Road, named for a well-known men's social club and rural refuge on the Amite River, established in the late 1890s.

The road was laid with gravel (an improvement) in 1938, but the area remained undeveloped—swamp forest and soybean fields—until the first subdivision, Audubon Estates, was built in 1965 on part of the property of a small mid-nineteenth century plantation, Audubon Farm. Today, settlement crowds the end of the road on the point of land where Bayou Manchac meets the Amite River. This peninsula, located across Manchac from the old site of Galveztown, has the lowest elevation in East Baton Rouge Parish.

Another nineteenth-century road wiggled from St. Gabriel to Galveztown—the old Spanish Way; a third paralleled the Mississippi through St. Gabriel on its way to Baton Rouge. But these dirt roads were muddy in wet weather, dusty and rutted otherwise, and difficult to travel.

Audubon Farms, also called Audubon Plantation, was one of two notable small properties located along the bayou during the mid-nineteenth century. The one-and-a-half story Greek Revival big house at Audubon was constructed about 1850, surrounded by fields of sugar that were later replanted in cotton. According to the 1888 Ascension Parish sales brochure (the same one that promoted Galveztown as a viable investment), Audubon's owners also ran a commercial lumber mill:

> Proof of [significant timber in this area] is furnished by the recent establishment of a paying enterprise at the Audubon Mills on the Bayou Manchac in East Baton Rouge about two miles below Hampton's Ferry. . . . Besides sawing cypress and other timber for mercantile or building purposes, band-saws and other improved machinery are employed in preparing for the northern market scrollwork, banisters, material for manufacture of furniture, railway cars, wagons and other vehicles. . . . Shipments are made direct from the mills to New Orleans on the steamboats plying in the Amite river and the Manchac twice or three times a week. . . . Fifty of such enterprises could find profitable employment for many years if scattered over this extensive section.

The other significant antebellum property was Santa Maria Plantation, originally a British land grant, located at the junction of Bayou Manchac and Ward's Creek. The land was unoccupied until the 1790s. Records indicate that a raised Creole cottage of local cypress and heart pine was constructed as the main house in 1834, and early maps identified a Manchac crossing there called King's Ferry. It connected Bullion Road from the Ascension Parish countryside with the growing commercial hub of Baton Rouge.

Like Audubon, Santa Maria initially grew sugarcane but converted to cotton sometime before 1901 when its sugar mill was also reconfigured as a cotton gin. In the 1940s, cotton was no longer the family enterprise, replaced by a popular dairy farm. Today, however, much of Santa Maria's original tract has been developed as a subdivision.

THE CIVIL WAR

When the Civil War erupted in 1861, although large plantations like Manchac and Woodstock lined the Mississippi, no large plantations or viable towns existed along the eighteen miles of Bayou Manchac. Furthermore, the bayou offered only partial navigability; it was therefore of little strategic interest to either the Union or the Confederate military during the long War between the States.

The War of the Rebellion: Official Records of the Union and Confederate Armies, an official transcript of letters, orders, and reports during the Civil War, mentions Bayou Manchac only occasionally. Other nearby places—Baton Rouge and Port Hudson, Doyal's Plantation on the Mississippi, the navigable Mississippi and Amite rivers—were of much greater significance, as was Pass Manchac, the strait between Lakes Pontchartrain and Maurepas.

References to the bayou in these records include such passing mention as the following:

> "We have information that the enemy is sending scouting parties out toward Bayou Manchac in the night of 100 men." (Confederate report, February 27, 1863)

> "Calvary and mounted infantry . . . proceed by way of Dutch Store, cross Bayou Manchac at Hampton's Ferry, thence to Amite. . . ." (Union army, September, 1863)

> "After attacking us yesterday, Scott fell back on the road leading to Bayou Manchac; the 11th New York pursued them to Hampton's Ferry where they crossed. The loss is about 80 men taken prisoner who were not able to mount, being sick." (Union army, August 6, 1864)

> "Bayou Manchac hereby announced as boundary between Baton Rouge and Port Hudson district and NO district until further notice." (Union army, August 20, 1864)

> "3rd Rhode Island cavalry policing the area from Manchac to Bonnet Carré." (Union army, May 22, 1865)

The bayou seems to have been relatively quiet during the Civil War, save an occasional skirmish or encounter. One of these took place in October 1864, when Confederate troops were captured at Hampton's Ferry over a period of several days:

> Sergeant J. H. Trigloan (First Texas Calvary) captured October 6, 1864, at Hampton's Ferry.

> Private J. W. Correy, First Louisiana Battalion Company, captured at Hampton's Ferry, Oct. 10, 1864.

> Private I. M. Mitchell, Third Louisiana Cavalry, captured at Hampton's Ferry, October 12, 1864.

One other minor Civil War encounter involved the capture of two flatboats from Bayou Manchac by soldiers of the 14th Maine regiment, seeking to cross the Amite. But this incident was overshadowed by the real action, which occurred when one of the boats, passing under a high bluff on the Amite just south of its confluence with Bayou Manchac, was ambushed in a nighttime attack by a small group of Louisiana cavalrymen.

With so little war activity around Bayou Manchac, owners of Santa Maria were surprised in the 1970s to discover several Civil War cannonballs lodged inside an ancient and diseased live oak that they were forced to cut down. It was a mystery why the cannonballs happened to be there—whether a previous resident had collected them from somewhere else and stashed them in the hollow trunk, or a Civil War skirmish had actually been fought close enough to Santa Maria that the oak tree had been hit.

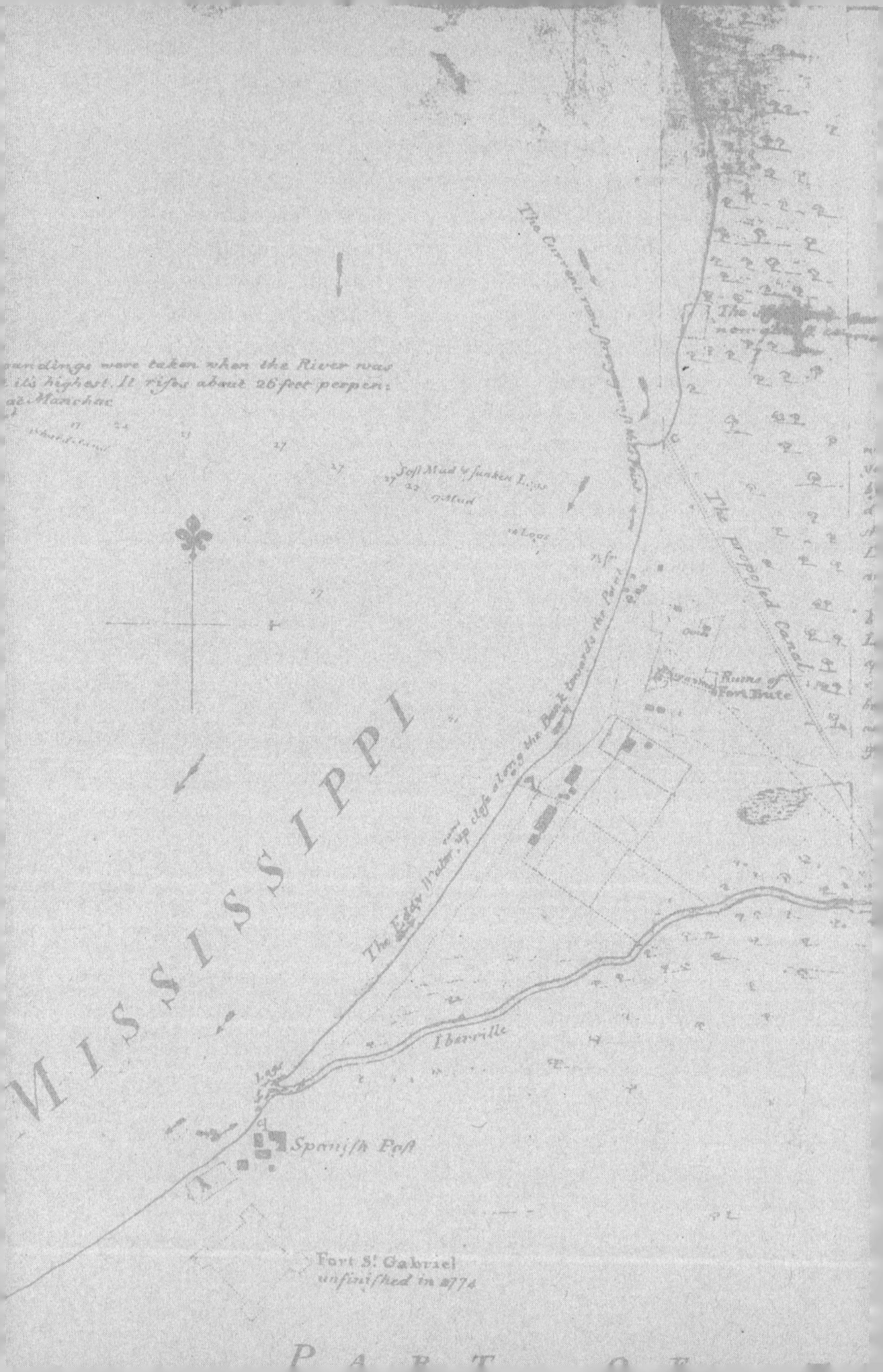
were taken when the River was
highest. It rises about 26 feet perpen:
at Manchac
Soft Mud & sunken Logs
MISSISSIPPI
The Edge of Water runs up close along the Bank towards the Point
The proposed Canal
Ruins of Fort Bute
Iberville
Spanish Post
Fort St. Gabriel
unfinished in 1774
P A R T O F

13 The Biggest Little Village on the Bayou

HAMPTON'S FERRY, THE PLACE consistently mentioned as the soldiers' crossing on Bayou Manchac, was the point where the dirt road between New Orleans and Baton Rouge met the bayou. The crossing was named for John Sidney Hampton, an East Baton Rouge Parish farmer who lived there and ran a cable ferry across the channel. The ferry was a primitive craft, a small, square wooden raft with railings that was pulled across on ropes or cables. But this simple transport in the wilderness became enough of a landmark to assure J. S. Hampton a footnote in history.

A crossing on such an important overland route was a natural spot to attract settlers, and after the Civil War, they began arriving. Among the first was Andrew Booth. The Confederate army veteran had returned to Baton Rouge after his capture at the Battle of Vicksburg and worked as a notary public until he decided to move his family to rural property on the upriver bank of Bayou Manchac.

Booth named his place Hope Villa because, he reputedly said, "this site holds great promise of success." He took charge of the ferry and established a store near the crossing. When traveler Samuel Lockett wrote of his visit to the area in 1869, he noted "Booth's Store or Hope Villa" as a destination; they were arguably one and the same.

By 1873, Hope Villa had a post office, which reflected official recognition of its existence. When Booth sold his store and moved to New Orleans in 1886, the small community survived him; it was on the map, both literally and figuratively. But Booth must still have believed in the area because he retained an interest in the property: a 1930 survey lists the owner of property on Bayou Manchac near the highway as A. B. Booth Jr.

The official location of Hope Villa, however, has become somewhat ambiguous. The Booth property on the East Baton Rouge Parish bank of the bayou was certainly Hope Villa, but the settlement that developed across the Manchac in Ascension Parish was also known as Hope Villa. Moreover, when Nathaniel Bishop had sought to travel Bayou Manchac in 1879, Mr. Walker told him about Hampton's Landing, not Hope Villa, although J. S. Hampton was no longer there and Hope Villa had already earned its post office.

Perhaps General O. A. Bullion's diverse activities helped muddle the geography. Octavius Alonzo Bullion was a newlywed when he came to clerk at Andrew Booth's store. He moved away, then returned in 1882 settling his family across the bayou, on the Ascension Parish side. Over time he became a local magnate and civic leader, earning the honorific of "General" although, according to local lore, he had turned down promotions from infantryman to officer on several occasions during the Civil War.

Bullion developed a small empire at Hope Villa, which included a general store that backed up to the bayou on the East Baton Rouge side and a cotton gin and mill, a moss press, and a blacksmith shop on the Ascension side. And, of course, he controlled the ferry. (His ferryman lived in a shanty near the bayou and charged five cents per crossing.)

The General also owned a large tract of farmland about two miles southwest of the bayou where he had built a house and grew cotton. This area became known as Bullion, the name affixed to the local railroad station, a stop on the Louisiana Railway and Navigation Company line, as well as to the road past Santa Maria.

The Ascension Branch Sugar Planter's Association touted Bullion's success in its 1888 brochure. It commended his store, which they located at Hampton's Ferry, "a beautiful little village situated on Bayou Manchac at the head of navigation directly opposite the village of Hope Villa." It also noted the General's success as a merchant "in the face of lively competition from merchants on the Hope Villa side of the Manchac."

The Old Hickory Masonic Lodge met on the second floor of Bullion's store, gathering during a full moon under the leadership of the General. Otherwise, this space served as a community center for church revivals, dances, and parties, while the first floor of the building held the store and post office, which remained in service until the 1930s.

O. A. Bullion served several terms in the state legislature as a representative from Ascension Parish—from 1892 to 1896 and again in the 1902–1903 session. He died in 1905 as a result of a gruesome accident in which his arm and leg were caught in a belt of his own cotton gin.

Hope Villa thrived because of its important location as the bayou crossing for the Jefferson Highway, the so-called model road and main roadway between New Orleans and Baton Rouge. The highway followed the Mississippi River north out of New Orleans to New River (now Geismar), then turned inland to Hope Villa and continued to Baton Rouge.

Since the trip from Hope Villa to Baton Rouge took several hours by horse and buggy, the little community (with a population of 125 in 1895) became the principal shopping center and trading post for the area. It boasted three stores clustered near the bayou, Bullion's cotton gin and blacksmith shop, a one-room schoolhouse, a resident doctor, and a chapter of the United Confederate Veterans of Louisiana. (They had accepted a donation in 1894 from the New Orleans chapter to repair damages caused by a Manchac flood.) A clutch of houses, some occupied by families whose landholdings dated to the late eighteenth century, lined the Jefferson Highway and the dirt roads that intersected it.

W. C. Abbott's family moved to Hope Villa in 1919, when residents identified each other by whether they "lived 'before you get to the bayou or over the bayou.'" In a self-published memoir about his life on Bayou Manchac, Abbott remembered the Hope Villa Mercantile Company store that rose on pilings over the bayou's bank. "We bought our groceries and most anything else we needed there," he wrote. But it was more fun to go behind the store, where an icehouse and turtle pen filled the sloping space on the bank. Ice shipped in from New Orleans was stored in huge packing boxes filled with rice chaff; the turtle enclosure held snappers caught by local fishermen until they were shipped to the New Orleans market for the Creole delicacy turtle soup. The ice arrived and the turtles departed by steamboat.

That was Hope Villa's other distinction: it was the terminus for the steamboat route from New Orleans. In keeping with the size of the Bayou Manchac channel, the boats were shallow-draft packets—"pocket editions of the Mississippi titans," as one traveler defined them—that carried both freight and passengers. They plied the lakes and the Amite River between New Orleans and Hope Villa, stopping at various landings along the way including Carthage, Clio, Whitehall, Maurepas, Tiger Bluff, Gunboat Bluff, Head of Island, French Settlement, Black Bayou, and Port Vincent.

Through the decades, many packets arrived at Hope Villa, including the *Amite,* the *Martine,* the *Katie,* the *Iberville,* and the *Jesse.* But the most popular and enduring of these was the propeller-driven steamboat *Alice*, first owned by the New Orleans–based Muncy Line of Delord Street and, later, by Muir and Fromherz, headquartered on the New Basin Canal in New Orleans.

The *Alice* began service to Bayou Manchac as early as 1868, as evidenced by letters from the property manager of Audubon Plantation, N. T. Kendall, to Audubon's owner, W. Frank Witherell. Witherell had recently purchased the property and had found both the Greek Revival cottage and its barn in great need of repair.

In a thin, even script, Kendall wrote to Witherell in New Orleans requesting supplies: "I wish you would please send me a three inch auger and handle so that I can commence the pig fence in front of the quarters at the earliest practicable moment." He lamented the threat of worms to their cotton crop and regretted the tardiness of his last letter, "which failed to go by the Alice."

In another letter, the property manager wrote, "I have much to say but the Alice will soon return." This remark implied that the little packet had passed on her inbound voyage and, when it had stopped to make a delivery at Audubon Farm, Kendall had requested a stop on her return. Or perhaps Kendall merely intended to stand on his dock and flag her down, a common custom.

Surviving handwritten receipts printed with the *Alice*'s logo reveal that W. Frank Witherell received many orders from New Orleans on the *Alice*. These included a trunk (its contents unspecified), a box of eggs, six sacks of seed cotton, two barrels of potatoes, a package of blankets, a roll of carpet, a barrel of apples, a coop of turkeys, and a box of buckwheat, among other necessities.

Display ads in area newspapers announced the *Alice*'s service. One prominent ad in 1877 read: "Old Reliable Muncy Line. Regular Amite River and Bayou Manchac, U.S. Mail. CSD Porter, Master; C. L. Brown, Clerk. Leaves New Orleans every Monday and Thursday at 3 o'clock." Its agents included "A. B. Booth, Hope Villa," who was, of course, pioneer Andrew Booth, whose store would have been a final stop.

Another ad from 1888 enthusiastically invited business: "The superb, elegant and fleet steamer. 'The People's Favorite' *Alice*. E. M. Robards Master; S. D. Porter Clerk, will leave New Orleans every Monday and Thursday at 2:30 PM for the Amite River, Bayou Manchac."

"Superb" and "elegant," however, were not the adjectives that Catharine Cole chose to describe the little boat. Catharine Cole was the pseudonym of intrepid *Times-Picayune* reporter Martha Field, who was a passenger aboard the *Alice* in February 1892. She reported her observations with cool journalistic objectivity. The *Alice*, she wrote, is "a square, dingy little boat more useful than beautiful . . . piled high with the delightful medley of freight usually sent to country folk. Machines, harrows, garden seed, rocking chairs, lamps, demijohns were packed roof high."* Field planted herself in "the tiny cabin reserved for ladies," which was furnished with several rocking chairs and a reading light, and chatted with the several other women aboard.

In another column, however, Field admitted that the *Alice* did have its charms, that it had been the area's "most faithful friend. How many billion feet of lumber the Alice has carried in that time only her owners know; but she has never failed the swamper nor the farmer and her record is a proud one, illustrating . . . the enterprise of honest and kind-hearted Mr. Alex Muir, her owner."

In a memoir about childhood visits to her grandparents' Santa Maria Plantation between 1892 and 1901, Gladys Mercer Morrill wrote of the *Alice* stopping to take on bales of cotton: "Our plantation was the last stop. . . . The boat could accommodate a few

* A demijohn is a large glass bottle used for transport or storage of liquids such as vinegar, wine, or liquors, usually protected by a jacket of wicker or raffia.

passengers and I once made the trip with my grandparents. There was always a group of people to meet the boat at the landings . . . like meeting the trains. . . ."

The mouth of Ward's Creek had been widened in 1895 to make a turning basin for steamboats. The young Gladys Mercer may have seen the little packet making the turnaround at the rear of her grandparents' property, its deck crowded with Hope Villa children. It was the captain's custom to invite them aboard to cruise the short turnaround with him, a sort of nineteenth-century amusement ride.

After 1880, packets shared the channel with barges servicing the burgeoning lumber industry. In the early twentieth century, however, steamboat traffic began to wane when overland carriers—especially railroads—became a more efficient way to carry commercial cargo.

Of course, the rickety, raftlike Hope Villa ferry continued operating until 1917 when the turn bridge replaced it. No doubt this was a cause for celebration, especially for the estimable Louis Babin. Though at the time he had not yet argued for making Manchac a link in the Intracoastal Waterway, Babin had used the cable ferry to cross his Hupmobile. On at least one occasion, the automobile rolled off the deck and into the bayou. It was fortunately rescued with a heavy rope and a team of mules kept handy for such incidents.

As automobile traffic increased, the turn bridge was replaced with the large new bridge in 1931, which allowed passage for commercial boats. But by then the back route was a pipe dream. Soon after, the Jefferson Highway lost its importance as well, replaced in 1935 with the new Airline Highway, a more direct route between New Orleans and Baton Rouge.

In the 1930s, with boat and overland traffic diminished, the closest railroad station located at Bullion, and its one-room schoolhouse replaced by a large new academy in nearby Oak Grove,

Hope Villa began to fade away. Among its last remembered landmarks was the Cedar Grove Casino, a rural nightclub at the foot of the bridge on the Ascension side, where a man stood outside ringing a bell to attract customers.

Today, Hope Villa remains as an official, black-lettered Department of Transportation sign on the East Baton Rouge side of the bayou and a two-block loop road named Hope Villa Drive on the Ascension side. A few houses from the nineteenth century survive along the old highway or are tucked among new subdivisions like reticent old aunts at a debutante party. And if the ghosts of O. A. Bullion and other families are still present in this suburb of Baton Rouge, their descendants have moved elsewhere.

Two miles east of Hope Villa, set back from the bayou, is a chain-link fence curling away from its posts. It encloses an overgrown patch of tall reedy grasses, blackberry thickets, a stand of saplings, and an untidy collection of white marble, puzzle-piece headstones. This is the "Myford Cemetery" as identified on detailed local maps.

The Gueldners and Meiferts, two German immigrant families, arrived at Hope Villa in the 1870s. Together, they homesteaded a thousand acres, farming cotton, corn, sugarcane, and sweet potatoes and raising sheep, hogs, and cattle. After a small dairy business failed, they turned to lumbering, selling a wood known as "baker's pine" for wood-fired baking ovens and white oak to manufacturers of barrel staves.

The little, hidden cemetery with its chipped, fallen headstones and half-legible inscriptions tells none of this tale. It merely attests to the existence of several generations in two families who lived and worked and died at Hope Villa. For the last recorded burial, that of Louise Sophie Meifert in 1949, there is no headstone. And none of her siblings, or any others of their generation or their descendants, are there amid the overgrowth. Like so many others, they abandoned Hope Villa.

For many years, unidentified people arrived infrequently to mow the grass, weed the plots, and repair the fence. But eventually, all the Meifert and Gueldner relatives who could remember those in the cemetery had died as well. No one has come to weed or clean the graveyard in decades. Only a misspelled dot on an obscure map acknowledges a family that, like Hope Villa, is almost forgotten.

One balmy day, I paddled Bayou Manchac where the Old Jefferson Highway bridge crosses the channel and tried to imagine Hope Villa as it might have been in its prime. But no traces remain of the old stores or the ferry landings, and the bayou banks are weedy with swamp privet or bare and littered. On the Ascension side next to the bridge is a green hillock, perhaps part of a former store site, the ferry landing, or the old bridge where a tethered white goat grazed forlornly. On the other side of the stream is a short private street of elegant, landscaped homes that back up to the bayou, separated from the channel by a line of fences that rises between their back doors and the channel.

West of the spindly-legged bridge, toward Ward's Creek, the bayou is densely scrimmed with trees, almost as if it were still wilderness. And the mouth of Ward's Creek remains anomalously wide from its days as the steamboat turnaround, spreading beneath the high terrace. If the *Alice* had happened along that day, she might have tooted authoritatively and hailed the tidy modern subdivision across the bayou.

I thought I could imagine the little packet at Hope Villa, the scuffed wood of her rustic main deck piled with freight, the low doorways inviting passengers into a spartan cabin. Or perhaps she would be animated with a crush of local children on the short joyride to the turnaround. And then I could envision her meandering back down the Manchac, weaving from bank to bank for scheduled and unscheduled stops under the long, green-leafed archways overhanging the channel. She would have been a cozy fit.

14 When the Woodmen Came

SOUTH OF THE CONVERGENCE, Bluff Swamp rises as a dense and beautiful tangle of bottomland hardwoods, thick with green textures and the complex scents of fertility and decay. It is also a museum of sorts, harboring a scattered handful of giant cypress trees, perhaps seven hundred years old. These prehistoric monuments soar skyward on tapering, moss-gray trunks, dwarfing the surrounding forest. They are miracle relics that somehow survived the golden age of lumbering.

Frank Bonifay and Jim Ragland, proprietors of the nonprofit Bluff Swamp Wildlife Refuge and Botanical Gardens, have no idea how the magnificent specimens survived the lumber boom of the late nineteenth and early twentieth centuries when old-growth virgin cypress was so efficiently harvested as to be almost wiped clean from the land for miles around. Perhaps these few trees were misshapen with burls or knots that made their wood too difficult to process, or maybe they were so inaccessible that even the best equipment could not reach them. Whatever the reason, the giants now live within a thirteen-hundred-acre preserve, links to prehistory and to Francisco Bouligny's description of the Isle of Orleans, when cypress grew "as thick as the hair on the head."

Cypress is a distinctive tree, called "the wood eternal" for its exceptional durability. It is immune to most insects and diseases, resistant to decay, and grows large and long-lived in soft mud where the trees' flaring trunks help balance their great height and weight. They are usually surrounded—and perhaps supported—by their sculptural protuberances called knees.

Cypress has been a choice construction material in south Louisiana since the first settlers. But until the 1880s, the great forests endured. Small operations like the Gueldners' and the Meiferts' as well as larger companies that cut and milled quantities of lumber were part of a traditional enterprise that had existed since 1716, when the first cypress mill was constructed near New Orleans.

Historically, cypress was difficult to harvest, not only because the trees were large and heavy but also because the swamp environment where they lived was so challenging. The best early method for harvesting cypress was a technique called girdling in which prime trees were selected during the dry season—late summer and fall—and a ring cut around their base to allow the sap to run out. By the following spring, when high water returned to the swamp, the selected trees were more easily cut and, having lost their sap, were light enough to be floated out of the swamp.

Because of the seasonality of harvesting cypress, sawmills operated only five months of the year. By the mid-nineteenth century, after the introduction of steam-powered saws, the process had become more efficient although the industry was still restricted by the difficulties of removing cypress from its habitat.

The federal Timber Act of 1876 changed the industry. It offered south Louisiana swamplands for sale at prices as low as twenty-five cents per acre, makin virgin cypress forests more attractive to investors. But the auction of over three million acres—a horror to some, even at that time—would have its rewards: the revenue would be dedicated to the building and maintenance of the state's levee system.

The low price of the land and the treasure that it held attracted large out-of-state companies, many of which had already cleared old-growth forests in their home regions. Cypress was the next big thing. Some companies bought land to log themselves; others hired local contractors. But they introduced to the region efficient modern equipment, which transformed logging the swamp and milling the lumber from small-scale seasonal operations to year-round businesses. At the height of the lumber boom, many mills in the area operated twenty-four hours a day for months at a time.

The revolutionary new machinery included steam-powered overhead-cableway skidders, pullboats, and small-gauge railroads. The skidders, patented in 1883, were capable of dragging logs from the swamp much more easily than men could; pullboats carried logs out down arrow-straight canals, which had to be dug specifically for their passage, as the boats couldn't maneuver in the naturally winding bayous. Rail lines to carry lumber through the swamp were set on beds elevated with dredged earth and overlaid with logs, support ties, and rails.

One of the largest cypress mills in the world, Lutcher and Moore Cypress Company, was headquartered in Lutcher on the Mississippi River, less than fifty miles from Bluff Swamp and Alligator Bayou. Its rival, Lyon Cypress Lumber Company of neighboring Garyville, was a main contractor for Bluff Swamp cypress, as were smaller companies like Opdenweyer Lumber Company on the Amite River and Natalbany Lumber Company of Louisiana, north of Lake Maurepas.

By the beginning of the twentieth century, the timber industry had surpassed agriculture as the largest economic force in the Amite River–Bayou Manchac area. It created new jobs for area residents and added to local populations by bringing in mill workers. The growth of the lumber industry also influenced the Corps of Engineers to desnag Bayou Manchac, the best route by which to ship logs to Amite River mills and beyond.

By the early 1920s, however, it was over: virgin cypress in the Spanish Lake and Bluff Swamp basins had been logged out.

Nevertheless, lumbering continued as other species of trees kept local mills whirring. A 1921 issue of the *Gonzales Weekly* newspaper described one such business, the Hope Veneer Company, with blatant boosterism: "[Their] post office address will be Hope Villa, Ascension Parish. The mill will be located at Bullion, in the same parish, on the Louisiana Railway and Navigation Company Road. . . . The timber to be cut consists of about 40,000,000 feet of magnolia, beech, gum, poplar, etc. It is said to be very large and fine and well suited for the manufacture of fine veneers."

The long-range effects of lumbering in the swamp and surrounding watershed proved much less positive than the industry's contribution to the local economy. The remarkable new harvesting machinery had introduced significant alterations to the landscape, which remained long after the cypress had been logged and the mills closed. The straight-line canals and ridge levees had redirected the natural water flow, creating noticeable disruption of the ecological balance in the cypress swamp and the lowland habitat around it.

The area was an extensive natural catchment consisting of Spanish Lake, Spanish Lake Swamp, Bluff Swamp, Alligator Bayou, and a network of small streams—Bayous Braud, Paul, Frog, and others. They lay at the lowest elevation in the watershed. By nature's arrangement, these lowlands held floodwaters and collected rain-

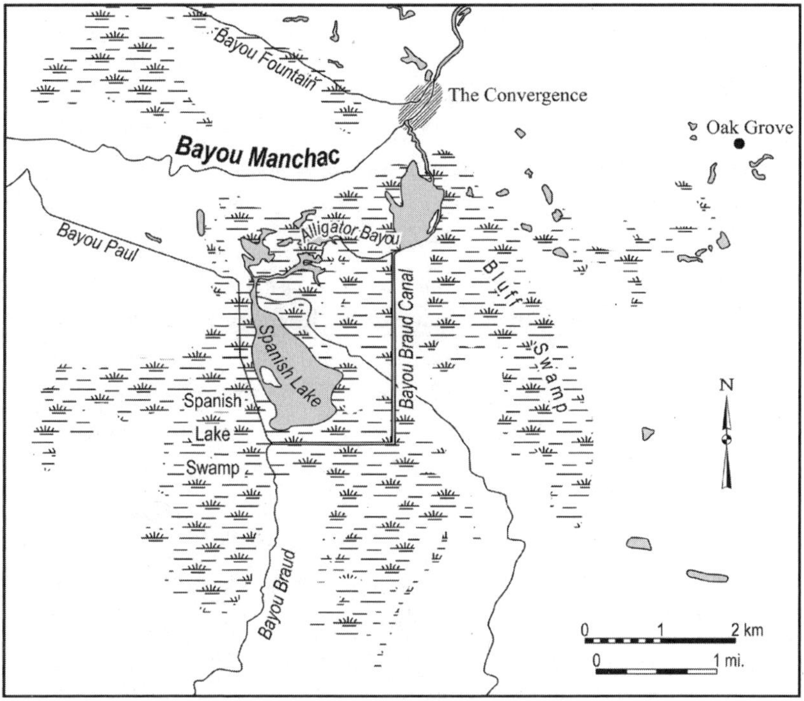

Spanish Lake basin

water that flowed in from Bayou Manchac and its sources and from Iberville and Ascension parishes to the rear.

The dynamic had long been understood. An 1842 report from the Secretary of War explained that "Bayou Crocodile [*sic*] . . . in high stages of water, before Manchac was closed, carried the river water into Lac Espagnol, overflowing lands almost to New River. The same lands are now drained by it into Manchac."

Philip Winfree, the advocate of damming Bayou Manchac at the Mississippi, had accurately described the Spanish Lake–Bluff Swamp area in his 1850 letter to *DeBow's Review:* "Between the

Mississippi and the Amite on the Bayou Manchac there is a deep swamp, part of which called the Big Swamp or Devil's Swamp, is at times twenty feet deep in water. . . ." That would be during spring flooding. And Lieutenant Hezlep's 1868 report also described the area: "The Crocodile [Alligator] Bayou is the most important [tributary of Bayou Manchac], being the outlet of Spanish Lake which receives all the drainage from Bayous Paul and Brand [Braud]." The natural hydrology of the system was abundantly clear.

Despite the changes wrought by the logging industry, the hydrology of the thirteen-thousand-acre expanse of swamplands and bottomland hardwoods remained predictable. In the 1950s, however, the Ascension Parish government, responding to citizens' concerns, raised the road that parallels Bayou Manchac, to create a protective levee between the basin and the bayou. Control gates were installed in the levee at the entrances of Frog Bayou and Alligator Bayou. The Bayou Braud canal was dredged and leveed along the southeast portion of the basin to funnel water away from Spanish Lake Swamp directly to Alligator Bayou and Manchac.

This reengineering protected lands around the basin but disturbed the balance required by swamp habitat. It altered or stymied natural water flow between and within the swamps: insufficient water arrived when it was needed, or too much came and remained standing. This created a destructive pattern that began to affect the health of indigenous plants and animals.

This threat to the landscape was, in part, the motivation for the creation of the Bluff Swamp Wildlife Refuge, which has zealously pursued a mission to save as much habitat as possible. But the proprietors, also wanting to educate the public about the natural history of Manchac and the Spanish Lake basin as a way to rally support, have created attractions.

The most popular of these is a fenced, ten-acre pen of fearsome alligators that loll on the banks of man-made ponds or float motionless and submerged, save a snout end and a raised forehead that give warning of their presence. Visitors accompanied by guides stroll through the pen. This has sometimes created the impression that the reptiles are stuffed or animatronic, like those at a theme park. But such skepticism is quickly dispelled when one of the large reptiles rouses itself and trundles quite lifelike across the dike, yawning to reveal its prominent teeth.

Less dramatic but equally interesting is a deep tank constructed to resemble a local pond. Filled with a variety of indigenous animals—garfish, turtles, and carp on my most recent visit—it offers visitors the opportunity to see what lives unseen in the tea-colored waters around the preserve and in Bayou Manchac.

The best adventure is a cruise aboard a small, covered raft that looks remarkably like photos of the old Hope Villa ferry, but with bench seats and a motor. The boat putters along Alligator Bayou and into the basin wilderness, offering unexpected sights of uncaged alligators lolling about, raccoons, nutria, migrating and native birds, vibrant splashes of wildflowers in spring, and drifts of young cypress crowned in rusty-olive leaves in fall. This voyage is also the best way to reach the towering old-growth cypress trees and to see the starkly beautiful expanse called Cypress Flats, a landscape of tall silhouettes rising from the never-ebbing mirror lake that killed them.

Since the invasion of overhead skidders and pullboats, most of what remains in the swamp is second growth. Still, given time and a benevolent environment, the swamp will thrive. How it should be restored, however, has remained a matter of debate. The most radical idealists hope to limit all development in the lowlands and return the large, historic catchment to wilderness. This vision, of course, clashes with development and political reality.

Nevertheless, the handful of seven-hundred-year-old cypress remain, trees that Iberville paddled near, and a testament to what the area once was. They are as deserving of protection as any historic monument but can survive only if the swamp in which they live remains healthy.

And that depends on what area residents decide.

15 Prehistory Redux

THOUGH THE OLD-GROWTH CYPRESS of Bluff Swamp are venerable and impressive, they are not the most ancient treasures found along Bayou Manchac.

In October 1932, several carloads of students dressed in waders and bathing suits and wielding picks and spades arrived on the south bank of the bayou, just downstream from the railroad bridge near Hope Villa. The students might have resembled fraternity pranksters on an outing, save for their distinguished chaperone. Dr. Henry Howe, revered professor and director of the LSU School of Geology, had brought his class to excavate the bayou in search of mastodons.

Mastodons were large, hairy vegetarian ancestors of the modern elephant that lived during the Ice Age, well before the Mississippi River took its current course and long before the formation of Bayou Manchac. Paleontologists believe that mastodons were killed off approximately ten thousand years ago, either by climatic changes or overhunting by early man. But at one time, they must have been relatively common, for Bayou Manchac is just one of eight sites in Louisiana that have yielded evidence of the woolly mammals.

The evidence that the students pursued had been discovered when eight-year-old Justin Cooper gashed his toe while playing in the bayou. Someone dove into the water to identify the cause of his injury and found a large tooth—three inches by six inches, according to a contemporary newspaper account. The exotic object was delivered to paleontologists at LSU, who certified that it was the tooth of a mastodon.

The team of students waded in the chilly water for several hours digging in the clay bed for other relics. They had reason to be hopeful, since Justin's mother had reported that her father had found a mastodon jawbone with two teeth in it at the very same spot thirty years before.

Howe's field trip was highly productive. By the time the students completed their excavations, they had pried a trove of fossils from the bed of the bayou—bits of skeleton, a knee joint thirteen inches thick, large molars, pieces of broken tusk, and sections of jawbones, all deeply embedded beneath three layers of clay. "Probably months or even years will be required to remove the specimens," the professor told the reporter who first documented the excitement for the local newspaper.

The mastodon find soon faded from the headlines, of course, and Justin Cooper retreated to an ordinary boyhood, though he bore a noble scar with an enviable tale attached. But he had made a lasting contribution to paleontology, as evidenced by a certain explicative text still in the archive collection of the LSU Museum of Natural Science. The citation for a certain large tooth reads: "Juvenile mandible, [from the] Rancholabrean epoch." Its listed collectors are "Cooper, Justin; Howe, Henry V."

16 A Quiet Stream and Modern Development

ALTHOUGH MASTODON FOSSILS lurked in the bed of the bayou, by the 1930s Manchac had grown quiet. Steamboats and barges stacked high with cypress logs no longer plied the channel, and the dream of the back route was effectively lost. A 1935 newspaper article reported that "slow moving houseboats and fishing skiffs are now about the only craft that move along the bayou."

It was a peaceful place where little had changed. In 1948, another journalist noted that "Manchac is little more than a lagoon which supplies some commercial fish, some game fish and some spots for relaxation and recreation."

The onetime international boundary had come to be appreciated for its natural bounty by boys with bamboo poles and commercial fishermen. The latter pulled great schools of buffalo fish, catfish, gaspergou, bass, freshwater carp, sacalait, bream, and garfish from its waters. And according to some old-timers who no doubt indulged in the usual hyperbole of fishermen, the garfish could measure seven or eight feet in length.

But the most commercially prized were buffalo fish, which swam up Manchac to Alligator Bayou and into Spanish Lake to spawn. Their run was a spring phenomenon that often occurred after a big rain, giving such downpours the local name "Buffalo rains." Bentley McKay lived on Bayou Manchac in the 1930s and 1940s and remembered them. He also watched expert fishermen gig buffalo fish with iron spears, deftly launching the points at the end of a thirty-foot rope. "The best of them could pierce a forty-pound fish swimming twenty or thirty feet away," McKay reminisced. One especially legendary fisherman was said to be able to hand-net or gig almost a ton of fish in a day.

However the buffalo fish were caught, when they were not sold commercially, fishermen hauled them home to be cleaned, cut into chunky fillets that were heavily salted, and carefully laid outdoors on wire lines to dry. If the smell of this local food staple made no lasting impression, its taste lingered in McKay's memory. "Considerably like dried codfish," he recalled.

Les Broussard was one of the commercial fishermen. He and his father trolled in a thirty-foot cypress skiff that they launched near the Hope Villa bridge, dragging a large homemade hoop net. They fashioned their nets during the quiet winter months, painstakingly knotting one-eighth-inch cotton twine with needles, then strengthening the homemade webwork by dipping it in a tank of boiling tar—a necessary precaution, explained Broussard, because a flopping, fighting mess of large buffalo fish or freshwater carp ensnared together could cut their way out of weaker threads. "We filled the boat twice with eighteen nets-full each time."

The Broussard men sold their catch to a wholesale market on the Amite River, where it was cleaned and sent to St. Louis. "People around St. Louis apparently ate a lot of buffalo fish," mused Broussard, which made their diet very similar to those who lived around Bayou Manchac.

The levee and flood-control gates erected along Manchac at Alligator and Frog bayous in the 1950s effectively curtailed the migration of buffalo fish, beginning the demise of commercial fishing along the bayou. But recreational fishing continued, even as pollution began to infect the water.

Hunters and trappers were also common along the bayou, as Ben Kleinpeter has indicated. One of them, Raymond Babin, lived along Bayou Fountain and went to "the dry part of Manchac," the first segment above the Convergence, to hunt and trap. There he bagged his limit in small game and sometimes encountered a trio of local women moss pickers. They were called the Three Maggies, he remembered, who collected the gray beardlike growths from tree limbs and sold it to area moss gins.* Cleaned moss, black and wiry like horsehair, was historically used as padding and stuffing in furniture, saddle pads, and packing material. The Three Maggies pursued moss collecting until the 1960s, Babin believed, when polyurethane and other synthetics universally replaced moss as a filler and put them out of business.

Even today, however, Manchac's remaining natural bounty attracts a few enterprising locals. There are scattered fishermen and a few frog giggers who come with their lights to prowl the bayou on chilly March nights. And at least one local entrepreneur has developed a small business in turtles, which he traps and sells.

One day, as a local story goes, this bayou entrepreneur was tending his traps when developers of a new subdivision spied him on their property—an unknown character engaged in what appeared to be a questionable activity. They confronted the man, who confessed to running his turtle traps.

Soon enough they'd made a deal: the developers offered the bayou resident the run of their tract, to trap every turtle he could find and

* Spanish moss, an epiphyte, is neither a parasite nor harmful to the trees on which it grows. A member of the pineapple family, it draws nourishment from the air.

clear the land before buyers came to inspect lots. Over several days, he had snared two hundred turtles to sell on the volatile turtle market, and the owners had taken a giant step in transforming their primitive bayou acreage into a properly civilized development.

17 Spirit of Place

AS I CAME TO KNOW the bayou, I decided that the best way to fathom its spirit of place would be to rent a cabin and spend some time there. I located a rustic camp, ambient with warped doors and bumpy floors, set back only twenty yards from the stream. The living area was raised some eight feet above the ground in a gesture of flood proofing, although over the years someone had enclosed the space below as a large ground-floor room. The new owners, weekenders themselves, had never known the bayou for big floods like the one caused by Tropical Storm Allison in 2001. They had begun refurbishing the upstairs living area, the first improvements the camp had seen in decades.

My retreat, located in Ascension Parish on a short road that parallels the bayou, was one among a string of dwellings—trailers, other rustic camps, a representative number of suburban-style houses, and primary residences, all of which overlook the Manchac.

On my first weekend a neighbor wandered over to see what a stranger might be doing there. She told me that she and her husband had moved to the bayou in 1974, when theirs was one of only three places along the road and had no phone service. But they treasured the isolation, and her husband loved the beauty of it, the natural history. He could name every tree along the bayou, every bird that perched in their yard, she said proudly.

Later the current road was laid out for a new subdivision, but it never progressed past the blueprint stage although other subdivisions had subsequently grown up all around them. She waved an animated arm inland, away from the waterway. "They've cleared out the pastures behind us and developed everything," she lamented, exaggerating only slightly.

My cabin lacked a raised, tree-shaded deck like the Caillouets', offering only an expanse of second-floor windows fogged with age that afforded a gauzy view through the trees to the stream. In August, the far bank of the bayou was a wild place, overgrown with shrubby swamp privet that rose against a backdrop of soaring trees, their limbs draped with hanging moss and long, trailing tendrils of pepper and rattan vines. I found myself looking out at a place removed from civilization.

In January, I peered through the same windows at the nearly bare winter landscape. The dense shrubs and vines across the bayou were almost leafless, silhouetted, revealing several trim, white-frame farmhouses and a pasture dotted with tractors and farm equipment. I smiled at the deception of August.

One late-summer afternoon, I sat on the narrow slab under the second floor overhang that passed for my porch, absorbing the quiet, utterly green landscape. The air rested thick and heavy, filled with the buzzing of crickets like a symphony of electrical short circuits. Suddenly, the leaves became animated, rustling wildly, the sky lowered in a charcoal hue, and thunder rumbled ominously and rolled across the sky above me, amplified by the open landscape. Shocks of lightning, wild and fiery, flashed so sharply that they transformed the sudden sheets of rain into a curtain of silver dashes blasting to the ground. And then, as quickly as it had come, the summer storm raced away, leaving the long grass glistening and the duckweed on the bayou swirled into a green abstract artwork. And the mosquitoes returned.

In the winter, the same landscape became a monoscape of gray, taupe, and tan; even the duckweed disappeared from the coffee-brown water. But a flurry of birds—cardinals and sparrows, finches, cormorants—cawed and chirped noisily, some providing darting flashes of color among the bare branches. On the mottled bark of a sycamore, the bright red head of a woodpecker, nearly glowing against the subdued landscape, hammered incessantly.

If the landscape changed, the bank of the bayou remained a slick of heavy red-brown clay, the color of a well-circulated penny, in every season. This clay could have been the rich material base for the pottery crafted by natives at the Kleinpeter site or used for bricks made by early settlers. However useful it may have been, the clay was almost impossible to scrape from my shoe soles after I had mucked through it launching my canoe.

The bulky green canoe and several paddles came with the camp and offered the opportunity to see the bayou from the inside, independently and at leisure. I sometimes brought a friend, and we paddled east toward the Amite or west toward the Convergence. We scraped the canoe up a narrow coulee to its end to see what was there, churned through the wide water at Ward's Creek, and beached on an empty bank beneath a ridge, scrambling up to see what was on the land. We encountered the occasional snake, a family of raccoons, a guidebook's worth of birds, but passed few other boaters.

And when I returned from such outings, the red clay of the bayou clung to my shoes, and I carried a little bit of Manchac's spirit of place home with me.

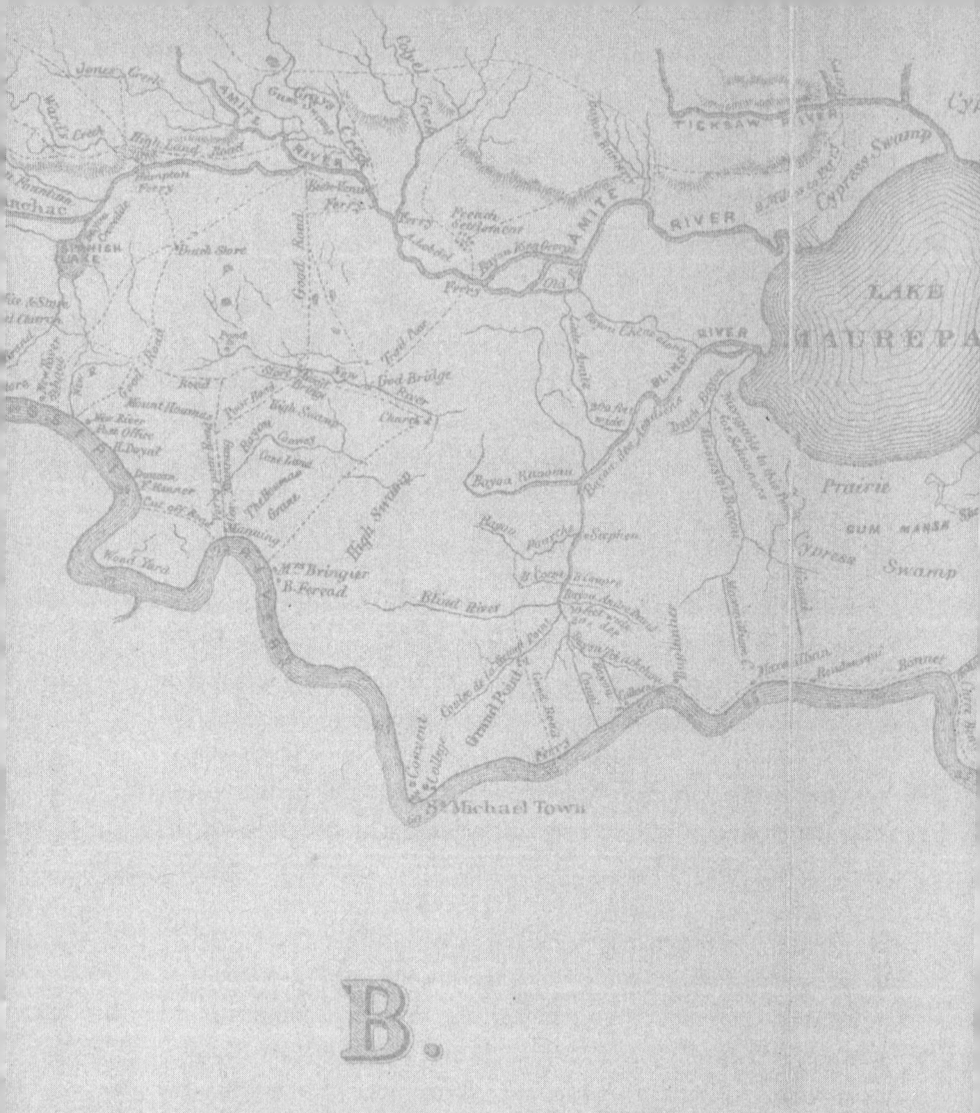

B.

t from Dep.t of the Gulf Maps No. 48 & 49.

Prepared under direction of

t. Brig. Gen.l M.D. Mc.ALESTER, Maj. of Engs.

accompany his Report on his Examination and Survey of

PASS & BAYOU MANCHAC & AMITE RIVER.

SCALE *of Miles*

6 0 6 12 18 *Miles.*

18 Contemporary Struggles . . . and Possibilities

UNTIL THE 1950S, life along Bayou Manchac remained rural and slow-paced. The banks were sparsely settled except for the cluster of houses and remnant businesses at Hope Villa. In most places, thick stands of forest studded the land and, where trees had been cleared, pasture and soybean fields rolled inland. In May 1953, when a regional newspaper reported dramatic flooding on the Amite River, the story offered sympathy for residents of Port Vincent on the Amite but mentioned Manchac only in passing. "Water is spilling over to some farm and pasture land that was inundated in the severe storms earlier this month," the article noted casually. The periodic episodes of backwater flooding disturbed few people along Bayou Manchac.

The bayou served its historic function within the complex but balanced watershed around it. On the East Baton Rouge side, rainwater that drained into Ward's Creek and Bayou Fountain flowed into Manchac and then east to the Amite River. If the water was excessive, it might also spill westward to Alligator Bayou and into the Spanish Lake basin.

When spring floods caused the Amite to backflow into Bayou Manchac, it could rise high enough to enter Bayou Fountain and Ward's Creek and even into the dry segment of Manchac. It could also drain into the Spanish Lake basin, filling the expansive tupelo-cypress swamp and bottomland hardwoods ringed by higher ground. Eventually, the water in the basin equalized with water in Bayou Manchac; then, as flooding ceased, the flow reversed, returning to Manchac and out to the Amite River.

This changed in the 1950s, however, as progress and new development began to affect the balance of the watershed. But as these changes took place, no one considered how "improvements" in one area might ultimately create problems in another. Nor did anyone think how these changes might ultimately affect Bayou Manchac.

Significant change to the bayou first resulted from enormous growth in the city of Baton Rouge. It had been originally settled on a bluff overlooking the Mississippi River, where Iberville first spied a reddened tree trunk separating the tribal lands of two native groups. He noted it in his journal as "le baton rouge," the red stick.

Here was the first high ground along the Mississippi above the Gulf of Mexico and a logical site for settlement. The French established a concession there, and though it failed in less than ten years, in 1763 the British built a fort and a cluster of huts nearby, the first permanent settlement. It grew slightly as a Spanish outpost and then as a bastion of the short-lived West Florida Republic.

After Louisiana became a state in 1812, however, the American Baton Rouge began to expand—slowly at first and then more robustly after it was designated the state capital and became a thriving river port. The 1850 census counted a population of slightly more than 3,900; a century later, it had become a city of 125,000.

Older sections of Baton Rouge were built on the extensive high ground that reached inland from the river. But as the population increased and city limits expanded in the twentieth century, homes, businesses, schools, and streets were constructed in lower, less-well-drained areas to the east. Some of these were located within the watersheds of Bayou Fountain and Ward's Creek and happened to fall within the one-hundred-year floodplain.

To alert developers and potential buyers to the inherent risks of the floodplain, the National Flood Insurance Act, passed in 1968, required that land elevations be made public. But the lure of undeveloped land at reasonable prices stoked burgeoning growth, replacing vast open tracts and dense forests. Green space that had previously absorbed rainfall (an average of fifty-seven inches annually) was transformed into hard, nonabsorbent surfaces—roofs, driveways, asphalt roads, parking lots—so that rainwater pooled, failing to run off quickly or to be absorbed into the ground.

In the early nineteenth century, Philip Winfree had offered an accurate description of Manchac's dynamic: "[Bayou Manchac] is the general drain of about one half of the parish of East Baton Rouge, of all Iberville on the same [bank], and of forty or fifty thousand acres of land in the parish of Ascension," he wrote. Perhaps this observation should have been taped to the office wall of every local developer and politician in the late 1950s, to remind them of the interrelatedness of the three parishes with the bayou, and to refresh their memories periodically about the concept of a watershed.

But this was of little concern.

Each vigorous new development in the areas surrounding Bayou Fountain and Ward's Creek replaced open land and exacerbated drainage problems. Normal hard rains flooded properties, and residents became annoyed, then angry.

Bryan Harmon is the current assistant director of the Department of Public Works for East Baton Rouge Parish and deals with such matters. "In the 1950s the push was drainage," he acknowledged. State money was available to help fund it.

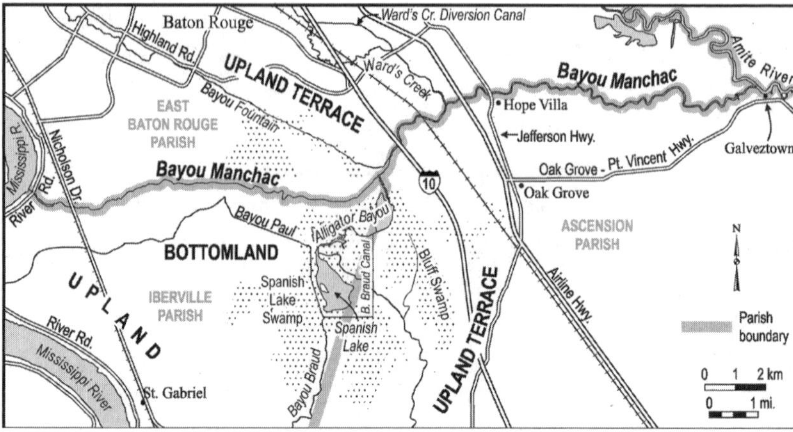

In periods of flooding, Bayou Manchac receives backflow from the Amite River, and waters from the surrounding Upland Terrace flow into the bottomland of the Spanish Lake basin.

In 1952, political leaders from Ascension and Iberville parishes approached their counterparts in East Baton Rouge to suggest finding a collaborative strategy for dealing with the increased flooding along Bayou Manchac. But the rapidly growing metropolis saw no reason to be influenced by its then-sleepy, rural neighbors. In response, it planned a parishwide public works program for drainage improvement, which was finally completed in 1965. Among the projects was a diversion canal built for Ward's Creek, a new, straight channel to funnel water more speedily into Manchac. Sections of Ward's Creek and Bayou Fountain were also lined with concrete to help whisk water away more quickly.

But drainage problems persisted and the expanding city continued to consider new ideas to solve them. Following the flood of 1973, Mayor-President W. W. Dumas suggested construction of a pumping station and holding pond at the junction of Bayou Manchac and the Mississippi. The idea was to push the water flowing down from East Baton Rouge Parish into Manchac in a westward direction, to an eight-hundred-acre man-made reservoir, then pump it over the levee. This system could also introduce fresh water from the Mississippi into Bayou Manchac when that stream was low. However, the newspaper account that reported the mayor's proposal did not mention that the first segment of the bayou was always "low" or, more accurately, dry; filling it with water might cause problems for nearby residents who lived primarily in Iberville Parish.

Dumas's proposal died without action, in part because the cost of pumping water uphill, against the natural lay of the bayou bed, would have been prohibitive. Also, clearing and dredging the dry segment and keeping it cleared would incur ongoing costs. But the proposal had revealed the parish's decidedly parochial attitude regarding the bayou.

"Bayou Manchac is the key to drainage in the southern part of the parish," said Ray Burgess, director of public works, in 1979. He made no mention of the bayou's history and seemed not to notice that the channel was bounded by two other parishes. Few residents of East Baton Rouge Parish lived along Manchac.

The first attempt by East Baton Rouge Parish to bring a more thoughtful approach to drainage came in 1990 after continued episodes of flooding. To mitigate the effects of new development within the parish, a new regulation mandated that a cubic yard of holding capacity be created for every cubic yard of construction. This required builders to dig a pond, a tank, or a lake to contain water equal to that which would otherwise run off. Moreover, these man-made holding areas could be designed to control how quickly water would be released.

Soon after, another regulation required structures built in specified low areas to be raised above ground to a certain level, based on the measure of the hundred-year floodplain. While this would not diminish runoff, it would at least decrease the number of properties that flooded.

Both measures proved wise because, by the year 2000, the population of East Baton Rouge Parish exceeded 400,000, and more than 40 percent of development was built in the floodplain.

But it was not only the city that had spilled over from high ground that caused problems along Bayou Manchac. After World War II, when increasing numbers of people began to enjoy the luxury of leisure time, the sleepy bayous of south Louisiana became appealing destinations as weekend retreats, Bayou Manchac among them.

In 1954, there were seventy-two camps sprinkled along the banks of both Manchac and the Amite. By 1972, the number had exploded to over six hundred. Furthermore, during the late 1970s, several gracious subdivisions of primary residences had been built along the Ascension bank of the bayou.

Clearly, as the number of people and structures along Manchac increased, so did the potential for flood damage. If the flood of 1953 had brought water only into pastures and fields, subsequent floods of 1973, 1977, and 1979 caused by backwater flooding created much more extensive damage to camps and homes along the bayou. These were followed by floods in 1980, 1983, 1990, 1994, and 1995, when an even greater number of properties lined the banks.

As camps and homes appeared along the Manchac, new residents also settled in Ascension and Iberville parishes, along the back rim of Spanish Lake and the Bluff Swamp basin and in the lowlands closer to the Mississippi River. Land grants in the area dated back to the Spanish colonial period, but early settlers had sensibly clustered along the natural high ridges. Many newcomers, however, built on low land or ran cattle in pastures that often held water. Like their neighbors across the bayou, however, these residents insisted on better drainage and more protection from flooding.

Backwater flooding from the Amite River was the historical and primary source of high water in the Spanish Lake basin. Occasional flooding also occurred due to high water flowing east from the Mississippi River. To protect the residents and regulate flow from the Amite backwater, the Ascension Parish government had raised the old gravel road along Bayou Manchac and installed the gates at Alligator and Frog bayous. Over the years, however, these measures proved inadequate.

Additionally, to protect the residents from waters coming from the south and west near the Mississippi River, Iberville Parish constructed leveed canals directing the flow into Alligator Bayou and, ultimately, into Bayou Manchac.

Each of these improvements protected a certain area but diverted the natural flow of water within the basin. And all of them contributed to the changes that had been begun during the great lumbering era.

The three parishes that bordered Bayou Manchac acted as if the Corps of Engineers were ultimately responsible for the bayou. But the Corps' primary duty, as it had shown historically, was to control flooding, when it could be justified according to a practical cost/benefit ratio. Ordinary maintenance of the bayou fell to the parishes.

This situation—not unusual for waterways that separate political jurisdictions—created a standoff. No one parish wanted to take responsibility. "The Corps must deal with the conflicting agendas of the three parishes that are bounded by Bayou Manchac," recalled Barton Rogers, a former project director for the Corps in the Manchac area. "And each of them," he noted pointedly, "has a different definition of 'what's in the best interest of the public.'"

Meanwhile, as parish leaders focused on concerns within their own boundaries, Manchac flowed on, utterly oblivious to political boundaries.

On June 6, 2001, a heavy, insistent rain began to fall. It lasted for four days and dumped twenty-one inches of water—almost two feet—directly on Bayou Manchac, Baton Rouge, and the surrounding area. Meteorologists called it Tropical Storm Allison.

At first, Pam and C. R. Caillouet were undisturbed. "After living here for almost ten years and flooding about once a year," Pam shrugged, "we had a feel for how high the crest would be, based on how much rain was coming down and where." They could predict high water by watching to see when the bayou began flowing backward from the Amite. They could also access United States Geological Survey gauge readings on the Amite that are transmitted by satellite to a public Web site.

In 1993, the Caillouets had watched as five and a half feet of water rose under their house, within two feet of the living area. They had stacked vulnerable items on cinder blocks under the house—lawnmowers, tillers, wheelbarrows, and such—and moved everything else from the yard to the top of their outdoor staircase or into their boat. They had parked their cars down the road on high ground that never flooded and tied their canoe to the banister, ensuring that they could paddle out to the cars or visit neighbors while floodwater still covered the landscape.

But Allison was different. "It was actually three seven-inch rain events," observed Bryan Harmon, and it brought a new kind of flooding to Bayou Manchac.

This time, long before the flow in the bayou reversed from high water on the Amite, water rose higher than the Caillouets had ever seen it. "We heard thumps and thuds and splashes under the house," Pam recalled. Their mower and tiller, which they had balanced on a five-foot woodpile, floated off, knocking into the support posts. A clothes dryer, tucked away under the house and destined for charity, bumped noisily against the floor joists and floated away. "There was nothing we could do at that point but watch it all go."

Engineers called it headwater flooding, water rushing down from the high terrace of Baton Rouge. For the first time in documented history, rain falling directly into Manchac and waters from Bayou Fountain and Ward's Creek combined to cause a catastrophic flood, before the Amite crested and backed up into the bayou.

At the height of the flooding, water rose six and a half feet, coming just inches beneath the floor joists of the Caillouets' log A-frame. The landscape was under water to every horizon, erasing landmarks such as their rural mailbox, planted on a sturdy post into the high ground near the road. It submerged completely. Trees looked shrunken as water hid their trunks to the height of a man's head. A house nearby that had never been raised disappeared to the tops of its windows. Masses of fire ants, with stingers intact, floated by in the swift brown water, attaching to anything solid—the side of a house, a canoe, a tree.

Just as the first flooding subsided, the expected crest of the Amite arrived and the backward flow began. The Caillouets, who had never feared water entering their home during their decade on the bayou, began to contemplate how they could possibly move furniture from their living area, more than seven feet above the ground, up the narrow interior stairway to their second floor.

And then, on June 11, the waters stopped rising and began to slowly recede. A week later, Bayou Manchac had returned to its banks.

The calamitous flooding caused by Tropical Storm Allison was a "200 to 300 year event," meteorologists and the Corps of Engineers agreed—a flood that should happen only every two or three centuries. The last time such catastrophic weather would likely have occurred was when the region was wilderness.

This event, however, brought huge quantities of rainfall to engorge the bayou, which was also partially blocked by downed trees that no jurisdiction had removed. And as the heavy rain falling in Baton Rouge spilled swiftly downhill through the chutes of Ward's Creek and Bayou Fountain, the holding ponds designed to control runoff had been overwhelmed.

"The [ponds'] holding capacity was good for the first seven inches," Bryan Harmon acknowledged. "After that, it was beyond the saturation point." The volume of water was so great, in fact, that the storage ponds remained full for a month after the floodwaters subsided.

Ward's Creek raged, topping its banks within the city and temporarily flooded the interstate highway. As its waters crashed into Manchac at the steamboat turnaround, terrified residents said it looked like Niagara Falls.

The water surged west and flowed back up Bayou Fountain; it also pushed against the gates at Alligator and Frog bayous. The fail-safe road-levee had been reinforced with sandbags, but the floodwaters overwhelmed the blockade and raced into Bluff Swamp across the lowlands. Kermit Braud, a native and longtime resident of Bluff Swamp, marveled that in seventy years he had never seen such high water near his doorstep.

Though such an event should occur only every two to three hundred years, the flooding caused by Tropical Storm Allison brought new concerns to the bayou. The initial floodwaters had not arrived from the expected source—backwater flooding from the Amite River. Nor did it occur because of one of the rare but documented instances of hurricanes causing backwater flooding from Lake

Maurepas up the Amite.* Neither could the flooding be specifically blamed on the logs and downed trees choking the channel. Tropical Storm Allison was the first incident in recorded history in which uncontrolled water from the high ground had overwhelmed the bayou by itself.

The extensive land development up the hill had surely exacerbated flooding, but other factors had contributed. The channel of Bayou Manchac was shallower than it had ever been, reduced from a depth of twenty-four feet in 1950 to a current average depth of approximately twelve feet, which significantly lessened the channel's capacity. The spectacular decrease resulted in part from the man-made barriers that thwarted the natural scouring action that had historically cleared the streambed. And the problem was exacerbated by quantities of dirt from construction sites along Bayou Fountain, Ward's Creek, and Manchac itself washing into the water and settling out in the channel.

Flooding has been a constant concern for residents along the bayou. But pollution is a newer problem. It has steadily intensified since the later years of the twentieth century, spoiling the bayou and creating an even deadlier hazard.

Sources of pollution vary, but one of the most egregious has been untreated or insufficiently treated sewage that flowed from overtaxed treatment facilities in East Baton Rouge Parish, where the rapid development not only strained drainage systems but also overwhelmed the parish's antiquated sanitary sewer system.

* One such incident, in September 1977, was caused by Hurricane Babe. More recently, scientists studying the devastating effects of Hurricane Katrina in 2005 noted that, had its path deviated as little as thirty-five miles to the west, immense volumes of water from Lake Maurepas could have pushed up the Amite and into Manchac and even up Ward's Creek and Bayou Fountain, with catastrophic results.

"Mainly during major rain events," conceded Harmon of the public works authority, "sanitation overflows are a problem."

In 1975, an extensive fish kill along Manchac was specifically traced to a sewage overflow. At that time, M. B. Watson, a biologist with the Louisiana Wildlife and Fisheries Commission, acknowledged that "Bayou Manchac is of marginal water quality most of the time." When asked why, he responded matter-of-factly that it was because of what drained into it.

Ten years later, an archaeological team arrived at the bayou to excavate a site thought promising for evidence of a native village dating from the time of European contact. It might also have contained remnants of a small, little-known British fort established soon after Fort Bute. But "excavators could not dive the bayou," revealed the official report of the Louisiana Division of Archaeology. The cause: raw sewage from the treatment plant on Ward's Creek.

An official sampling of bayou water in 2000 by the Louisiana Department of Environmental Quality (DEQ) produced further distressing results: water quality did not meet standards "for fish and wildlife propagation or primary and secondary contact recreation." In other words, the bayou was unhealthy for fish and other animals and a risky place for humans to fish, boat, or swim. Of course, the report maintained optimistically, the bayou could support a diversity of fish—white crappie, bluegill, channel and blue catfish, and possibly yellow bullhead, freshwater drum, bowfin, carp, buffalo fish, and gar. But only if pollution were successfully controlled.

The DEQ's report also detailed sources of the bayou's degradation beyond the faulty sewage treatment plants. These included improperly treated wastewater from industrial and commercial sites, construction and bank runoff that brought soil and contaminant particles, and household and garden chemicals. Rain carried all of these into the city's drainage network and into the Bayou Fountain and Ward's Creek watersheds, to be washed down into Manchac.

Trash and litter follow the same pathway. Most of the bottles, Styrofoam, and papers that bob against downed trees in the bayou and settle on the slick mud banks began as discards on city streets that were flushed into the city's drainage system. The junked household appliances, building materials, and such, however, originate right along the banks of Manchac itself.

Fortunately, there is promise of curtailing some of the pollution in Bayou Manchac: a federal consent decree has ordered East Baton Rouge Parish to thoroughly upgrade and renovate its sewage system. By December 2014, the parish will have rehabilitated sewer lines and made treatment plants efficient, creating a system with sufficient capacity. In the interim, Bryan Harmon assured me, there will be incremental improvement, progress "on an ongoing basis." The project will ultimately cost more than $1 billion.

But even if sewage from the upland terrace no longer pollutes the bayou after 2014, no one has yet sufficiently addressed the problem of sewage from the exploding population in Ascension Parish. The bustling development along Manchac and in the Spanish Lake basin has heavily taxed their sewage treatment facilities. And, in some areas, residents rely solely on individual septic tanks or small, inadequate community facilities that leak and malfunction, sending pollution into Bayou Manchac. However, that parish has recently begun to require more effective treatment systems in new construction and remodeling projects to modernize their septic systems.

The bayou has faced other threats as well. In 1998 traffic engineers planned an interstate bypass around Baton Rouge, calling for a broad new highway to run through the Bayou Manchac and Spanish Lake basin area. Few citizens realized where the winding stream was or why a broad roadway there would matter. Only when a small cadre of environmentalists protested publicly, fearing further devastation to the basin, did the planners rethink their proposal.

Then, in the aftermath of Tropical Storm Allison and after an extended hiatus, Ascension Parish invited the Corps of Engineers to help clear and desnag the bayou. But Manchac residents expressed concern that the contract language called for "clearing from bank to bank," a phrase that suggested that the scenic beauty remaining along parts of the channel might be destroyed. Further, removing trees that grew at the bayou's edge or pitched slightly over the channel would only increase bank erosion.

After a series of heated discussions froze the timetable for the project, the Ascension Parish government unilaterally cleared and desnagged the bayou. "They floated in a huge barge, about two-thirds as wide as the bayou, pushed by a small tug," Pam Caillouet related. She had been among the most wary residents. "When they finished, the barge was piled high with tree trunks . . . like a mountain moving past my deck."

The channel was successfully cleared, at least until trees inevitably began to fall again into the stream. East Baton Rouge Parish had not taken part in the project.

19 Where Are the Rose-Colored Glasses?

IN THE BEST OF ALL possible worlds, history is always referenced, conservation and development evolve together, and a man-made vision of place is tempered by the realities of nature.

This state of harmony is exceptional, however, and definitely not what has evolved for Bayou Manchac.

It is tempting to imagine that, once Manchac had survived the dream of the back route, local jurisdictions would have mutually ensured that it remained a valued natural asset—protected for its history, its scenic beauty, and its recreational importance. Instead, the little waterway fell victim to degradation that has undermined it perhaps as much as if commercial development had taken place.

There have been occasional glimmers of hope that Manchac might garner some attention, but these promises have not yet translated into effective action. In 1991, for example, the Amite River Basin Commission was created, an umbrella nonprofit agency whose multiparish purview was to "facilitate and coordinate flood mitigation efforts among all levels of government." Its founding principle was idealistic, but the organization's broad territory (reaching from Lake Maurepas north to the state of Mississippi) and its low level of funding has made it weaker than any of the entities it serves.

The commission's current focus is a six-to-nine-year project underwritten by federal, state, and local entities to create a diversion canal on a tributary of the upper Amite River. The construction targets the upper Amite; a tangential effect may be a slight decrease (just over a foot at best) in the level of backwater flow that enters Bayou Manchac during a one-hundred-year flood. But it is not a project intended to protect the bayou.

Dietmar Rietschier, the commission's director, understands the need for improvement in the Manchac watershed and acknowledges that it could be achieved with cooperation among the three parishes that border the bayou. "The parishes need to talk," Rietschier observed. "They share that watershed . . . but in the context of realpolitik," he admitted, "it's hard to do."

In early 2005, the Corps of Engineers announced a study on the feasibility of enhancing the entire Amite River floodplain. Its lofty goals, set in print, included restoring the ecosystem, improving habitats, and reducing flooding, pollution, erosion, and silting. But after Hurricane Katrina in August of that year, the Corps' attention and funding in south Louisiana was diverted. Only at the end of 2006 did a preliminary meeting for this project—scaled down—take place. And with the required study, hearings, and public input required on any Corps of Engineers project, whatever action that might result is four or five years in the future.

Representatives of government and land-use agencies convened an urban watershed conference in East Baton Rouge Parish in 2005. They gathered to discuss wetlands and greenbelts—how to identify surviving natural areas in the parish and determine what enhances or threatens them. The intent was to influence land development and improve flood protection and water quality. If the conference ever develops findings that can be translated into effective programs, Bayou Manchac and its residents might receive direct benefits.

Another alluring possibility for improving the future of the bayou is to showcase its rich history with an application for designation as a Maritime Heritage Landmark. Part of the National Historic Landmarks program within the National Park Service, this citation recognizes significant and historic canals and waterways that, in the language of the agency, "possess exceptional value or quality in illustrating or interpreting the history of the United States." The federal program helps preserve, protect, and promote such places.

Bayou Manchac, site of prehistoric native cultures and visits by Iberville and Bartram, host to British, Spanish, Acadians, and Germans, steamboats, loggers, and forty-pound buffalo fish, might be an excellent candidate for this designation.

Unfortunately, however, the bayou must first be in good condition to qualify.

And this raises the inevitable question: what will become of Bayou Manchac now?

Development will surely continue along the bayou's banks and in the surrounding areas. Perhaps it will do so with greater vision and wisdom. A plan to preserve the bayou could be devised that considers the interdependence of each part of its complex watershed and promotes collaboration. Bayou Manchac might become a regional priority.

These are the possibilities seen through rose-colored glasses. But if none of this comes to pass, Bayou Manchac will simply continue to wind through time. And, as it does, it will become more and more diminished until—ultimately—it will be suitable only for its ghosts.

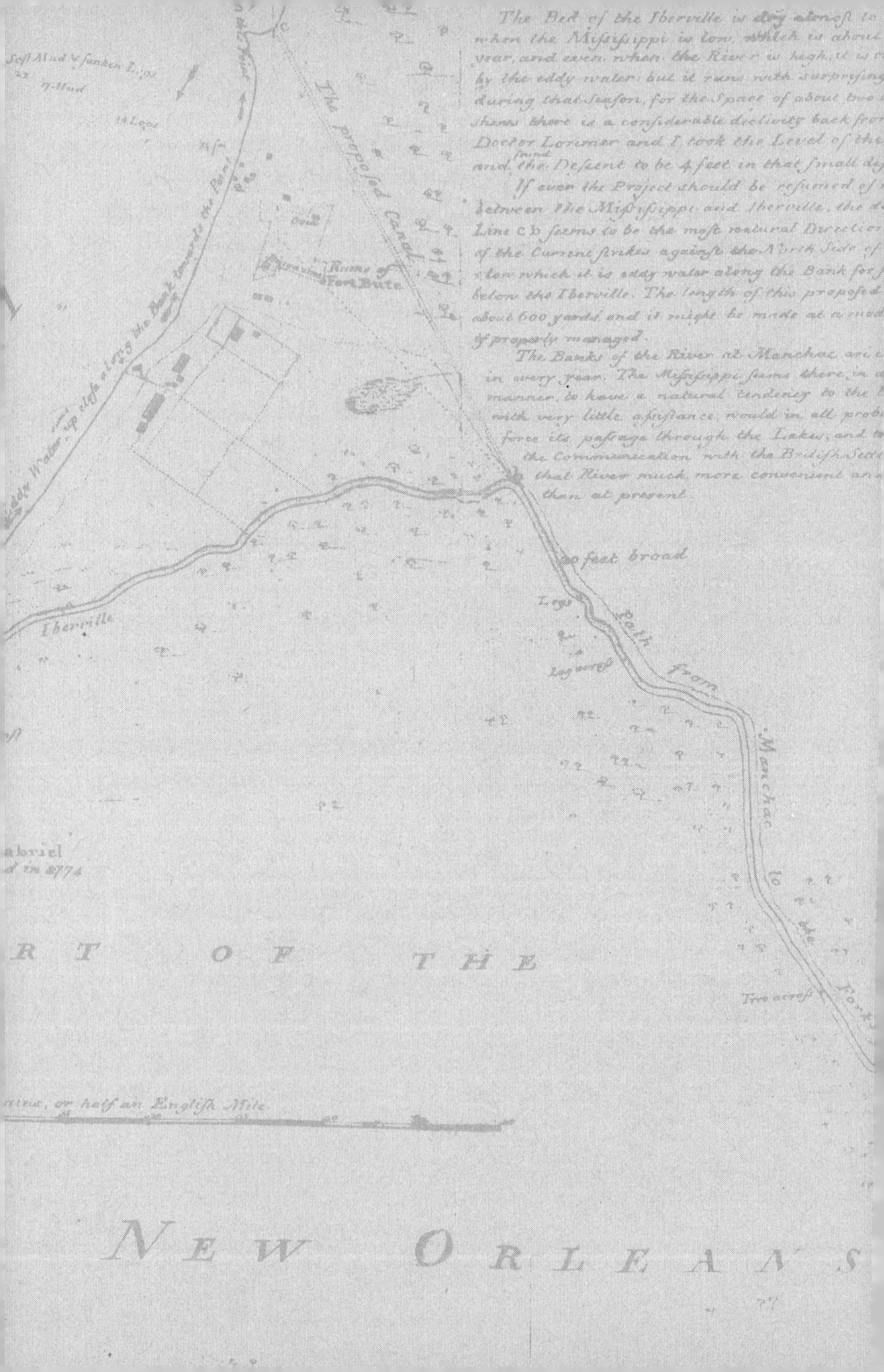
The Bed of the Iberville is dry almost to
when the Mississippi is low, which is about
year, and even when the River is high, it is
by the eddy water: but it runs with surprising
during that Season, for the Space of about two
shews there is a considerable declivity back from
Doctor Lorimer and I took the Level of the
and found the Descent to be 4 feet in that small di
If ever the Project should be resumed of
between the Mississippi and Iberville, the
Line C D seems to be the most natural Direction
of the Current strikes against the North Side of
elow which it is eddy water along the Bank for
below the Iberville. The length of this proposed
about 600 yards and it might be made at a mod
if properly managed.
The Banks of the River at Manchac are
in every year. The Mississippi seems there, in a
manner, to have a natural tendency to the
with very little assistance, would in all prob
force its passage through the Lakes, and to
the Communication with the British Settle
that River much more convenient and
than at present.
The proposed Canal
Ruins of Fort Bute
Iberville
20 feet broad
Logs
Path from Manchac to the Fork
Tree across
abriel
d in 1774
RT OF THE
NEW ORLEANS
ains, or half an English Mile

Notes

Native American cultural periods: Native American cultures in the Lower Mississippi Valley region are divided into the following periods:

250 B.C.–200 A.D.	Tchefuncte
200–400 A.D.	Marksville (first appearance of burial mounds)
400–1100 A.D.	Troyville/Coles Creek (earthen mounds at the Kleinpeter site most probably built during this period, used to support temples or ceremonial buildings)
1100–1450 A.D.	Plaquemine (introduction of farming)
1450–1700 A.D.	Mississippian
1600–Present	Historic period, after European contact. Muskogean linguistic group included the languages of Bayou Goula, Acolapissa, Mugulasha, and others.

Isle of Orleans: When France divided its territory in 1763, the Isle of Orleans became a Spanish territory. The following is a translation of the pertinent text from the Treaty of Paris defining its geography:

> In order to reestablish peace on solid and durable foundations, and to remove for ever all subject of dispute with regard to the limits of the British and French territories on the continent of America; it is agreed, that, for the future, the confines between the dominions of his Britannick Majesty and those of his Most Christian Majesty, in that part of the world, shall be fixed irrevocably by a line drawn along the middle of the River Mississippi, from its source to the river Iberville, and from thence, by a line drawn along the middle of this river, and the lakes Maurepas and Pontchartrain to the sea; and for this purpose, the Most Christian King cedes in full right, and guaranties to his Britannick Majesty the river and port of the Mobile, and every thing which he possesses, or ought to possess, on the left side of the river Mississippi, except the town of New Orleans and the island in which it is situated.

William Bartram: Bartram is recognized as America's first native-born naturalist/artist as well as the first to pen a journal of both scientific observation and personal experience. His lengthy southern journey in the late eighteenth century took him through the pristine environment of eight modern-day states: North and South Carolina, Georgia, Florida, Alabama, Mississippi, Louisiana, and Tennessee. The account of his adventure, known as Bartram's *Travels,* first published in 1791, includes his impressions, letters, travel notes, and natural history observations, and has been widely read. His path has been commemorated by the establishment of an informal series of routes known as the Bartram Trail, identified on the Bartram Trail Web site, http://www.bartramtrail.org.

Francisco Bouligny: Bouligny owned land in the Manchac area that he neglected or chose not to develop, as Gilbert C. Din reports in *Francisco Bouligny: A Bourbon Soldier in Spanish Louisiana:* "In

1792 Manchac residents complained in a memorial to the governor about absentee landowners, including Bouligny, who failed to maintain their levees, which harmed their properties. Governor Carondelet gave the absentee owners instructions to repair the levees by November or face a one-hundred-peso fine." At the turn of the century, Bouligny still owned abandoned lands at Manchac.

Los Isleños: In August 1777, the Spanish crown commanded the governor and commandant general of the Canary Islands to enlist seven hundred men for service in Louisiana. Preference was given to married men who would bring families to populate and stabilize the community. Mulattoes, gypsies, executioners, and butchers were prohibited from volunteering.

The Spanish government paid each traveler 45 reales upon agreeing to go to Louisiana, a half peso per day while awaiting departure, and 45 reales upon arrival. (A reale was one-eighth of a Spanish peso, or piastre, a coin that was sometimes cut into eight pieces, giving rise to the phrase "pieces of eight.") In 1778–79, over two thousand Canary Islanders left on ships for Spain's Louisiana colony, where they were dispersed to Galveztown and to three other south Louisiana settlements—New Iberia, Terre-aux-Boeuf, and Valenzuela.

William Darby: Darby's extensive travels in Louisiana in 1816 were recorded in his report, *The Geographic Description of the State of Louisiana,* which combined detailed observations of terrain and physical geography with knowledgeable commentary on natural history. His description of the forests of West Florida near the Mississippi River, including the upper bank of Manchac, is as revealing as that of William Bartram.

Darby noted "a thick forest, of which the most remarkable [species] are Liquidamber styraciflua (sweet gum), Pinus taeda (loblolly), Pinus rigida (pitch pine), Cupressus disticha, Ulmus aquatica (swamp elm), Acerrubrum (red flowering maple), Quercus tinctoria (black oak), Quercus alba, Quercus virens (live oak), Quercus acquatica (water oak), Quercus phellos (willow oak), Juglans amara (bitternut hickory), Juglans porcina (pignut hickory), Nyssa sylvatica (black gum), Nyssa acquatica (tupelo), Fraxinus tomentosa (red ash), Salix nigra, Celtis crassifolia (blackberry), Gladitsia triacanthos (honey locust), and Diospiros virginiana (persimmon).

"Also, the Liriodendrum tuplipfera of large size. And the laurel magnolia abounds over the whole extent of this plain."

A universal pest: Although William Darby was quite wrong about Galveztown's future prospects, he was much more accurate about one of the area's greatest afflictions: mosquitoes. Sixty species of the insect are native to the state, some directly responsible for the malaria that plagued Galveztown and other south Louisiana settlements. In fact, mosquitoes were historically among the most unpleasant aspects of life in the area. Darby waxed almost poetic in his exasperation with the pests: "The never ceasing hum of these creatures awakes the mind of the person exposed to their bite the most disagreeable sensations," he complained. "They are the insect hydra. . . ."

William Bartram had also commented on the fury of mosquitoes when he camped along the Amite River the night before he reached Bayou Manchac. Bartram was less lyrical but equally perturbed: "Fire and smoke were insufficient to expel the hosts of mosquitoes that invested our camp and kept us awake during the long and tedious night so the alligators had no chance of taking us napping."

An earlier traveler, John Pope, grumbled in 1792 that "for two days past we have been much harassed by mosquettoes." He noted that the natives, "who go about almost naked," had devised an effective strategy for coping with this nuisance while sleeping: "[They] construct an elevated bed of reeds which they suffumigate [build a smoky fire beneath] so as to banish insects of every description from their lodgements."

In 1906, Clifton Johnson traveled beyond the New Orleans area to go alligator hunting. He apparently enjoyed the adventure but wrote, "[M]osquitoes keep the population down; otherwise [local residents would] be overrun with hunters and fishermen."

These and others who visited south Louisiana through its three centuries of recorded history rarely failed to comment on the ubiquity and unpleasantness of the menace. And if they returned today, they would do so still.

Boundaries of the Louisiana Purchase: Text (translated) from the Treaty of San Ildefenso, 1800, regarding the definition of the transaction in North America: "the colony or province of Louisiana, with the same extent that it now has in the hands of Spain and that it had when France possessed it, and such as it ought to be according to the treaties subsequently concluded between Spain and other states."

From the Louisiana Purchase document, 1803:

> [France] doth hereby cede to the United States in the name of the French Republic for ever and in full Sovereignty the said territory with all its rights and appurtenances as fully and in the Same manner as they have been acquired by the French

> Republic. In the cession made by the preceding article are included the adjacent Islands belonging to Louisiana, all public lots and Squares, vacant lands and all public buildings, fortifications, barracks and other edifices which are not private property.

Jean Lafitte: In March 1812, Governor C. C. Claiborne declared that smuggling, especially of slaves, was illegal. He offered a reward of five hundred dollars to anyone who could deliver the privateer Jean Lafitte. Lafitte, his brother Pierre, and their band of followers based in Barataria Bay south of New Orleans, had successfully eluded Claiborne's efforts to capture them.

As the British made plans to attack New Orleans, two British officers approached Lafitte, hoping to entice him to join their invasion. Lafitte wrote Claiborne to inform him of the plot, even as the reward was on his head.

When Andrew Jackson arrived in New Orleans and discovered how unprepared the military was, he accepted Lafitte's offer to serve. Lafitte asked that he and his men be given "the opportunity of proving that, if they had infringed the revenue laws, yet none were more ready than they to defend the country and combat its enemies." Lafitte's men helped defend the Mississippi River forts and the New Orleans area. Some of the Baratarians, as Lafitte's band was called, served with Jackson in the battle itself.

Lafitte was remembered as a romantic figure and is the subject of numerous south Louisiana tales. One suggested that he had lived along Bayou Manchac near Bayou Fountain and buried money on the point there. Another said that Lafitte threw a golden bell in the water between Bayou Fountain and Alligator Bayou. Yet another declared that Lafitte's gold had fallen off a boat in the Amite River, near the site of Galveztown.

Notes from the late W. C. Abbott's memoir, *Bayou Manchac in the 1920s*:

> E. D. Dixon was the last person to run the Hope Villa Mercantile Company store, about the time the new bridge [1931] was built. The new road [to the bridge] needed part of the land where the store was located. Someone bought the building and rolled it a mile down the road, made it into a nightclub, but later it burned down.
>
> [There were] lots of boats in the bayou. Some houseboats were pulled by a launch. I remember one man stayed anchored for several years. [He lived on the houseboat.] We used to go to his boat and fish from it.

When he was a boy, Abbott recalled,

> our main interests in the bayou were swimming and pole fishing. We walked along the shore with a cane pole, a line, a hook, and a can of worms. [We] found a place just behind the old boathouse in front of our home where we could catch dozens of red sun perch. Another good place to fish near our home was a tangle of logs and stumps and roots from an ancient tree that had fallen into the bayou. Bluegill, bream, and goggle-eye perch gathered there to hide from garfish and bass. I didn't fish there much because it was easy to get your hook hung on a log. If you have only one or two hooks, you learn to fish in more open places.

Pam Caillouet's further recollections of Tropical Storm Allison:

> When it floods, objects of all sorts float down the bayou. After Allison, I was out in the canoe, chasing down some of our possessions that were floating away, when I saw a flash of red in the

water. I snagged it and hauled it into the boat. It was an afghan that someone had spent many hours crocheting; it needed only an edging to complete it. I hung it on the back deck for a while, hoping the owner would claim it, but no one ever did. I'll always wonder what its story was.

[Another memory] is of seeing the Salvation Army truck parked as [close to the bayou] as it could get, and each day a volunteer was there, handing out hot, home-cooked meals. We didn't partake, because we had enough provisions, plus we had electricity [even though their electric meter was totally submerged] and were able to cook. However, we heard repeatedly from several neighbors that the food was "absolutely delicious." The truck came [for] about two weeks. There were lots of folks without power and some who had been flooded, [so] the Salvation Army food wagon was a welcome sight.

Acknowledgments

This book might never have been written without the enthusiastic response of Mary Katherine Callaway, Director of LSU Press, who believed the story of the bayou should be told. It is a story inspired by the dedication of Pam and C. R. Caillouet and Jonathan Scott, who are Bayou Manchac's best friends. The Web site that Jonathan created, www.bayoumanchac.org, served as a source for information about the bayou long before I began my research.

In gathering material, I was grateful to have the patient and efficient staff at LSU's Hill Memorial Library, who were supportive in every way, and to have the wonderful resource of the Louisiana and Lower Mississippi Valley Collection. I also received significant help from The Historic New Orleans Collection and from Sam Hyde and Keith Finley at the Center for Southeast Louisiana Studies, Southeastern Louisiana University.

Individuals who shared information and help are always invaluable to a project like this. My thanks go to Pie Lanoux; Eugene LeBlanc; Ira Paul Babin; Dale Clary; Steve Fullen and Rebecca Saunders, LSU Museum of Natural Science; Pat and Glenn Cambre; Barton Rogers; Dennis Jones; Malcolm Shuman; Frank Bonifay; Clark Forrest; Joan and Jack McLaughlin; Beth Ann Naugher; Amy Grayson; Kermit Braud; David Broussard; Frank Bonifay;

Lowell Urbatsch; the late W. C. Abbott; Ben Kleinpeter; Les Broussard; Bentley McKay; Raymond Babin; Richard Condrey; Louis Capozolli; and Ward Bond.

I would be remiss if I failed to recognize the few previous writings focused on Bayou Manchac that reminded me in the midst of struggling to find material that a few others had also deemed writing about the bayou a worthwhile pursuit. These include "Louisiana's Mysterious Manchac" by Mattie Bell Cawthon Singletary (LSU master's thesis, 1931); "The Role of Water Transportation in the Settlement of Bayou Manchac and the Amite River," by John Reilly McKenna (LSU master's thesis, 1975); W. C. Abbott's memoir *Bayou Manchac in the 1920s*; and Leroy Willie's *Bayou Manchac and Its People*.

Special gratitude is also due to Ron Boudreaux, who remained a cheerful accomplice in gathering experiences along the bayou and who learned to canoe in the process. And, of course, to Joanna, Katie, and Jay, who inquired regularly about the book's progress.